BEHAVIORAL
AI

ROGAYEH TABRIZI, PhD

BEHAVIORAL AI

UNLEASH DECISION MAKING WITH DATA

WILEY

Copyright © 2025 by John Wiley & Sons, Inc. All rights reserved, including rights for text and data mining and training of artificial technologies or similar technologies.

Published by John Wiley & Sons, Inc., Hoboken, New Jersey.
Published simultaneously in Canada.

No part of this publication may be reproduced, stored in a retrieval system, or transmitted in any form or by any means, electronic, mechanical, photocopying, recording, scanning, or otherwise, except as permitted under Section 107 or 108 of the 1976 United States Copyright Act, without either the prior written permission of the Publisher, or authorization through payment of the appropriate per-copy fee to the Copyright Clearance Center, Inc., 222 Rosewood Drive, Danvers, MA 01923, (978) 750-8400, fax (978) 750-4470, or on the web at www.copyright.com. Requests to the Publisher for permission should be addressed to the Permissions Department, John Wiley & Sons, Inc., 111 River Street, Hoboken, NJ 07030, (201) 748-6011, fax (201) 748-6008, or online at http://www.wiley.com/go/permission.

Trademarks: Wiley and the Wiley logo are trademarks or registered trademarks of John Wiley & Sons, Inc. and/or its affiliates in the United States and other countries and may not be used without written permission. All other trademarks are the property of their respective owners. John Wiley & Sons, Inc. is not associated with any product or vendor mentioned in this book.

Limit of Liability/Disclaimer of Warranty: While the publisher and author have used their best efforts in preparing this book, they make no representations or warranties with respect to the accuracy or completeness of the contents of this book and specifically disclaim any implied warranties of merchantability or fitness for a particular purpose. No warranty may be created or extended by sales representatives or written sales materials. The advice and strategies contained herein may not be suitable for your situation. You should consult with a professional where appropriate. Further, readers should be aware that websites listed in this work may have changed or disappeared between when this work was written and when it is read. Neither the publisher nor authors shall be liable for any loss of profit or any other commercial damages, including but not limited to special, incidental, consequential, or other damages.

For general information on our other products and services or for technical support, please contact our Customer Care Department within the United States at (800) 762-2974, outside the United States at (317) 572-3993 or fax (317) 572-4002.

Wiley also publishes its books in a variety of electronic formats. Some content that appears in print may not be available in electronic formats. For more information about Wiley products, visit our web site at www.wiley.com.

Library of Congress Cataloging-in-Publication Data is Available:

ISBN 9781394196869 (Cloth)
ISBN 9781394196852 (ePub)
ISBN 9781394196845 (ePDF)

Cover Design: Wiley
Cover Images: © pressureUA /Getty Images and Wiley

SKY10094328_122324

To my parents, who planted the seed of knowledge in my mind and nurtured it, you are my universe and constant source of inspiration and strength.

To my team and all who cheered me on during this incredible journey, and who saw in me what I couldn't see in myself.

To all my teachers, especially the great ones, who sparked my curiosity and ignited my passion.

To the challenges that made me stronger.

Contents

Preface		*xi*
Chapter 1	Magic Happens at the Intersections	1
	Asking the Right Questions: Data, Intuition, and Strategy	3
	Simplifying the Complexity	6
	Connecting the Dots	8
	Uncovering Hidden Patterns: Models and Algorithms in Action	10
	Decoding Consumer Behavior: The Interplay of Psychology and Economics	11
	Empowering Behavioral Economics: The Synergy of Data Analytics, ML, and AI	14
	Crafting a Customer-Centric Paradigm: The Fusion of Technology and Behavioral Insights	17
Chapter 2	It Is All Connected: Behavioral Economics, Decision-Making, Biases, and Heuristics	19
	History and Origins of Behavioral Economics	20
	Early Days	21

	Entering Mainstream Economics	23
	Current Research and Practical Applications	26
	Back to the Beginning	30
	Psychology of Decision-Making	33
	Dual-Process Theories	33
	Heuristics and Biases	35
	Noise	39
	Prospect Theory	40
	Nudging	42
	Experimentation	46
	How It Works	47
	Uber and Experimentation	49
Chapter 3	Minimal Data, Maximal Impact: From Big Data to Minimum Viable Data	53
	How Much Data Are We Talking About? Lots and Lots	55
	You Do Not Need a Lot of Data to Get Started, You Need the MVD	58
	Asking the Right Questions, Again!	59
	Synthetic Data: What It Is and What It Isn't	63
	Survey Data to the Rescue	67
Chapter 4	Building Intelligence: AI and ML Essentials, Transforming Data into Intelligence	73
	Classical AI	76
	ML	78
	Deep Learning	82
	Generative AI	86
	Machine Intelligence and Biologically Inspired Models	90

Chapter 5	Real-World Impact: Harnessing AI and ML for Practical Solutions	95
	Unleashing the Full Potential of AI: Beyond the Hype	96
	Rethinking Segmentation: Beyond Demographics and Life Stages	97
	Uncovering Unexpected Customer Patterns	101
	Predicting Intent and Mapping Customer Journeys	104
	Overcoming Challenges in Predicting Customer Intent	106
	Predicting and Managing Returns	108
	The Power and Nuances of Recommendation Models	111
	Broadening Horizons: Beyond Category Killers	114
	Enhancing In-Store Experience with Recommendation Models	116
	Leveraging Propensity Models for Targeted Campaigns	118
	Personalized Pricing: Influencing Behaviors and Financial Outcomes	121
	Behavioral Economics in Personalized Pricing Strategies	124
	Forecasting: Understanding the Dynamics of Demand	127
	The Power of Forecasting and Optimization	131
	Transparent MMMs	132
	Ensembling Models for Enhanced Forecasting	133
	The Interplay of Demand Forecasting and Inventory Optimization	133
	Conclusion	135

Chapter 6	Decoding Complexity: Leveraging Systems Thinking in Modern Organizations	137
	Only a Wet Baby Likes Change: Loss Aversion + Status Quo Bias	140
	It Gets Better! Commitment Device, Peer Effect, and Sunk Cost Fallacy	145
	Conclusion	151
Chapter 7	Unlocking Scale: Overcoming Operational and Organizational Complexity in Scaling AI Projects	155
	Enablers of Success	156
	Communication and Intellectual Diversity	159
	Building Trust, Experimentation, and Adoption	164
	Interpretation Layers	166
	The Power of Experimentation	169
	Measuring ROI Through Experimentation	171
	Conclusion	173

Epilogue	*175*
Notes	*179*
Additional Reading	*189*
Bibliography	*197*
Acknowledgments	*205*
About the Author	*207*
Index	*209*

Preface

"I have no special talents. I am only passionately curious."
– Albert Einstein

Much of this book has been written during plane rides to and from client meetings and conferences and is filled with stories, examples, reflections, and considerations drawn from countless conversations with clients, executives, and practitioners.

In these pages, I aim to take you on a journey through the realms of data science, behavioral economics, and organizational complexity. Together, we'll explore how to harness the power of predictive models and AI to uncover hidden patterns in data, drive informed decision-making, and ultimately create value within your organization. By sharing real-world examples and case studies from various industries, my goal is to provide you with practical insights and strategies that you can readily apply to your own work.

We will discover how to effectively combine data-driven insights with an understanding of human behavior to overcome challenges and drive change within organizations. We will explore concepts such as game theory, cognitive biases, and behavioral economics, and how they can be combined with machine learning and AI models to enhance decision-making processes and improve customer experiences.

Navigating organizational complexities and overcoming silos can make it challenging to identify the right datasets, build high-quality models, and implement them effectively to achieve tangible results. Through the numerous stories and experiences shared in this book, I aim to provide you with actionable strategies to harness the power of your data. Additionally, you will gain a deeper understanding of the significance of addressing cognitive biases and fostering a collaborative, transparent environment to ensure the success of your initiatives.

Ultimately, this book aims to equip the practitioners and executives with the tools and knowledge needed for AI and behavioral economics to navigate the complexities of modern organizations, make data-driven decisions, and lead your team to success. Through the lessons and stories shared, I hope to inspire you to embrace innovation, challenge the status quo, and unlock the full potential of your organization.

My own story starts on a plane ride from Tehran to Vancouver. I grew up in Tabriz, a city in the northwest of Iran, at the intersection of many cultures: Turkish, Azari, Persian, Kurdish, among others. As a young kid, due to my father's job, we traveled and lived in many different cities where we often didn't speak the native language, and frequently moved in the middle of the school year.

Being competitive with a natural drive for excellence, combined with my love for mathematics, made me pay extra attention – not just to my teachers, but to everything around me – to make sense of things and stay a top student. I remember paying attention to what drove my classmates, recognizing similarities and differences across cultures and language barriers. Without realizing it, I was being trained in behavioral sciences, but it took me almost 20 years to connect the dots.

I arrived in Vancouver when I was 21 to pursue my master's in physics, specifically in string theory. I had fallen in love with a particular area of mathematics called algebra, group theory, and algebraic topology, which lends itself beautifully to string theory and particle physics. This might sound strange, but the beauty, elegance, and power that this area of mathematics has to simplify the complexities of our physical world have guided my thinking and approach all these years later in the work I do with Fortune 100 companies.

Shortly after starting my master's, I switched to particle physics. This was partly because, no matter how much I love math and theoretical physics, I need to see the physical manifestation of these fascinating theories – I need to see their impact or reality in practice. I want to experiment and see what happens. I call this my *impact bug*. The timing was also perfect: CERN was up and running, almost! I was one of the last groups of people who got to see the ATLAS detector open. It was a moment that I will never forget, one that changed my life and my relationship with the word *impossible*. I remember thinking that if we can build this machine that helps us understand what happened 10^{-16} seconds after the big bang, we can do anything!

I sometimes jokingly say, "Impossible doesn't quite occur to me!" and it's thanks to that moment. I also will never forget the first time I saw the Tier 3 facilities at CERN: a computer farm so vast that you couldn't see the end of it. That was my introduction to big data. The first real data we received after the initial collisions was 7 terabytes (TB), and it didn't even occur to me that it was big. It was just what it was! We had to figure out how to parallelize our "jobs" over however many cores and pray to God that we had caught all the bugs in our codes before submitting them!

There were many other lucky moments for me during my time at CERN. Getting to know the senior management at CERN was one, and having the opportunity to be the youngest member of the organizing committee of the First African School of Physics was another. I got to see the inner workings of one of the most sophisticated and complex scientific collaborations: 10 000 scientists from more than 100 countries coming together for 40 years to make a Nobel Prize–winning discovery. Only years later did I realize how I was being trained in systems thinking and life at 32° Fahrenheit, as described in *Loonshots* by Safi Bahcall,[1] by some of the best scientists in the world.

In *Loonshots*, Bahcall talks about phase transitions, where small changes in conditions can lead to dramatic shifts in behavior. This concept was evident in the way CERN managed to align diverse talents and resources to achieve groundbreaking discoveries. I also got to see how some of these very same physicists came together and asked a simple question, leading to a decade-long journey: How can we build foundational capabilities in fundamental physics in a continent, Africa, with massive potential? With careful planning, building relationships, and networks of the right people and institutions, I witnessed more than 700 graduate students participate in a three-week summer school in fundamental physics over 10 years. More than 70% of them ended up doing their PhDs and postdocs in Europe and North America, with 35% returning to their home countries. I got to see how a movement starts and spreads across a continent, much like a phase transition where initial efforts create a ripple effect, leading to significant and sustained impact.

All this and more led me to decide to study economics for my PhD. I wanted to delve into development economics. The reason was simple: How is it that we spent $20 billion and 40 years to build a machine that can answer what happened 10–16 seconds after the big bang, but spent $3 trillion over 40 years in international aid in Africa and still can't keep a polio vaccine cold enough to reach a village? I wanted to understand what we can do better, what we can do differently. The impact bug had hit me yet again.

Determined to pursue the economics of development, I started my PhD only to fall in love with game theory in the first semester. Game theory is the study of why people do what they do, the way they do it. It was the second most beautiful thing I had ever encountered after the standard model of particle physics. I was particularly fascinated by social and economic networks and how games of incomplete and asymmetric information unfold within these networks. I wanted to understand what drives behavior when individuals are influenced by those around them, or why a particular technology takes off rapidly in one society but never gains traction in another. What makes some people more influential than others, and what are the underlying conditions that lead to positive spillovers or strategic complementarities?

These questions led me to explore the interconnectedness of human behavior and economic outcomes, as beautifully articulated in *Social and Economic Networks* by Matthew Jackson,[2] one of my mentors and advisors. His work provided profound insights into how networks form, evolve, and influence economic activity, shaping my understanding of the complex web of interactions that drive development and innovation.

As much as game theory was different from particle physics – as if I only wished people would behave like atoms – I was still a model generation machine. It took me nearly three years to realize that I was undergoing a paradigm shift, trying to understand how an economist thinks and why it differs from the mindset of a physicist. Every week, I arrived at my supervisor's office with my shiny new model. After patiently listening to me explain it, he'd invariably ask, "But what is the question?!" or "What is the intuition behind the model?" I painfully anticipated this question every time. It wasn't that I didn't have an answer; I just didn't understand the question! I had a model with a clear set of assumptions explaining a particular dynamic, equilibrium state, or evolution in the behavior of my "agents." Very physicist of me!

It took me three years to understand what he meant because, slowly, I was developing an *intuition*. I was learning, ever so subtly, to question the questions, to listen, and to search for the underlying, hidden assumptions that I was making without quite realizing it. I was also learning to identify them in others' arguments and papers. I began to make connections between seemingly unrelated, sometimes contradictory, facts. It was starting to become natural to me, much like how I played around in my head with group theory and algebra. I was able to tap into an area of math that felt natural to me and use the art of asking questions to peel back the layers of the onion one by one, find the right question, and backward engineer from there. It was first principle thinking in action.

This transformation was crucial, as it enabled me to bridge the gap between abstract models and real-world applications. It wasn't just about creating elegant models anymore; it was about ensuring these models had a solid foundation in reality and could provide genuine insights into human behavior and economic dynamics. I also realized that in physics, I was trained to optimize: to find the best, most efficient way to solve a given problem. In economics, however, the question is never given. Our job is to first find the right question and then solve it. It was magical to sit at the intersection of these disciplines and seamlessly – much to my advisors' surprise – go back and forth between them. This skill has become my most important asset in solving complex problems today. It's a skill I see sorely lacking in many data scientists trained in physics, computer science, or engineering.

But transitioning from academia to the real world wasn't simple, especially with my background as a theoretical physicist, an experimental physicist, and then an economist. The flexibility to apply different methods and tap into tools across multiple disciplines, coupled with the ability to think more fluidly and flexibly, plus many years spent in university, made me a strange outsider when I decided to leave academia and not pursue a tenure-track position.

Yet, this multidisciplinary training became my superpower. It enabled me to approach problems from unique angles, ask questions others might overlook, and combine insights from various fields to create innovative solutions. When I decided to move beyond the academic world, I brought with me a tool kit rich in diverse methods and a mindset geared toward continuous learning and adaptation. This journey has shaped my ability to navigate and thrive in the complex landscapes of today's data-driven world.

The first year was the hardest: like many with ties to universities in Silicon Valley, such as Stanford, I started my career at a startup in the Valley. Shortly after, I moved back to Vancouver to work as a data science specialist for a unicorn in Canada. I was the most senior data scientist in the company, and my life couldn't have been more different from walking the hallways of the economics department at Stanford or sitting in the cafeteria at CERN. Nothing in my education had prepared me for the real world, especially for talking to "normal people" (as I jokingly used to say). No one understood what I said, why I suggested a particular method or approach, or why I was asking for more data. Anything I did, and how fast I did it, felt like magic to my colleagues.

So, I had a brilliant idea: leave a well-paying job and do a bit of consulting to figure out what I wanted to do when I grew up. I asked a few friends to introduce me to executives in different industries. That's how Theory+Practice was born, and an unexpected phenomenon occurred almost immediately: The first client I signed was a billion-dollar retailer, the second was a $2 billion e-commerce company, and the third was the world's largest logistics and express transportation company – all within the first six months of the company. The second year was even more surprising, as I signed with the largest retailer in the world, followed by several other major Fortune 500 retailers and financial institutions, and then the largest consumer packaged goods and supply chain companies in the world.

I had experienced very steep learning curves when transitioning from theoretical physics to experimental physics, and again when moving from a master's in particle physics to a PhD in economics without any background in economics. But this wasn't a learning curve – it was a learning wall!

I was fascinated by enterprise and all the incredible complexity of these environments. Looking at problems from the freshest, most outsider perspective possible, I believe this was one of the key reasons behind the traction we gained. I couldn't understand why we needed to wait for all the data to be cleaned and organized before starting to answer the most important and strategic questions. I remember jokingly telling executives that if we had waited to clean 20 TB of data at CERN, there would be no Nobel Prize, no discovery! I wanted data and a lot of it, and I wasn't going to wait. That's how – paradoxically – I started using the phrase *minimum viable data*. I knew that by applying proper methodologies from the intersection of different disciplines, we could build high-quality models to test various hypotheses.

I also attracted an incredibly multidisciplinary team of physicists, economists, engineers, and computer scientists whose curiosity and hard work pushed the boundaries of what everyone around us claimed to be impossible. We loved tackling seemingly impossible problems.

I have dedicated most of my adult life to studying ways to detect patterns in vast quantities of data, whether it was searching for the signatures of the Higgs boson among oceans of other particles or signals for the underlying factors driving particular behaviors. Being trained at the intersection of physics, economics, and behavioral sciences is humbling, yet it has enabled me to cross-pollinate and realize that there are many ways behavioral science can benefit from methodologies used in physics and computer science, and vice versa.

By illustrating the practical applications and lessons learned in the following chapters, I aim to show how economic intuition and behavioral economic methods can become incredibly useful tools. These approaches not only help to decode the complex patterns hidden within vast datasets but also drive meaningful and impactful decisions. Through this exploration, I hope to provide valuable insights into how we can leverage these interdisciplinary methods to solve some of the most challenging problems in today's data-driven world.

CHAPTER

1

Magic Happens at the Intersections

'I would rather have questions that cannot be answered than answers that cannot be questioned."
– Richard P. Feynman

In September 2022, I found myself under the spotlight at a renowned retail conference, the first in-person conference since COVID-19. The pandemic had accelerated the digital and data transformation efforts of many large enterprises by more than five years. Big data, artificial intelligence (AI), and machine learning (ML) were among some of the most important sessions because many executives believed that they could use these advanced methodologies and predictive models to understand what their customers' wants and needs were, and how they could serve them better by providing more personalized products, prices, and marketing. However, they were wondering how they can use vast amounts of data to predict how customers' demands have changed in order to optimize their productions and shipments to minimize excess inventory as well as out-of-stock challenges.

I shared how so many of my conversations with executives across finance, insurance, retail, and consumer packaged goods all start with statements like, "We have so much data and we do not know what to do with it." I gave examples of using AI and ML models from our recent projects during the pandemic. Then, I addressed the elephant in the room: consumer preferences have changed, and it's more crucial than ever not to rely solely on deterministic models that fail to capture these dynamics.

I paused on a slide with the title "Magic Happens at the Intersections of Data, AI, and Behavioral Economics" to ensure the message was clear. Each of these topics alone is complex and the subject of many books and discussions. Individuals earn their PhDs in AI or behavioral economics after years of study and research. And here I had 45 minutes to talk about the magic at the intersection of these disciplines!

AI and ML encompass the creation of sophisticated models and algorithms designed to execute tasks traditionally requiring human intelligence. These tasks span a wide range of capabilities, including reasoning, learning, perception, problem-solving, language comprehension, and even creativity. At their core, these models are driven by data, learning to identify patterns and make informed decisions with minimal human intervention. One of the most remarkable aspects of AI and ML is their ability to adapt to new circumstances and continuously improve over time. This adaptability not only enhances their accuracy and efficiency but also opens the door to innovative applications across various industries, transforming how we approach complex problems and decision-making processes.

Behavioral economics, however, blends insights from psychology with economic theory to understand how people actually make choices, which often deviate from the predictions of traditional economic models based on rational decision-making. Human decision-making is multifaceted, influenced by emotional, psychological, social, and contextual factors. Yet, so many of the engagement interactions with customers – such as discounts, email campaigns, or customer service calls – are based on a "one-size-fits-all" approach and often face challenges in influencing the nuanced behaviors of consumers and customers.

Human decisions are rarely purely rational; they are interwoven with emotions, cultural norms, past experiences, and even seemingly illogical biases. The static structures that customers interact with – whether online or offline – struggle to keep pace with the dynamism of consumer preferences that evolve with societal trends, technological advancements, and shifts in cultural values. By integrating the power of AI and ML with the deep insights of behavioral economics, organizations can craft more personalized and effective strategies. This fusion enables businesses not only to anticipate and respond to customer needs more accurately but also to create meaningful and engaging experiences that drive loyalty and satisfaction.

Asking the Right Questions: Data, Intuition, and Strategy

Many of the companies and executives whom I have worked with are keen on extracting "actionable insights" from their data to support their intuitions and strategies. However, many have previously collaborated with various consulting firms, which often failed to deliver meaningful insights, leaving them skeptical about new projects. One executive in a Fortune

500 company referred to a collaboration with us as a "last effort to make sense of the data," and another mentioned that prior experiments and A/B testing had not expedited their quarterly goals. These companies, despite having revenues exceeding $1 billion and vast amounts of data, struggled with leveraging their data effectively.

The common issue is usually the lack of clarity about their assumptions and questions. Although data can easily answer straightforward questions about product popularity or support sophisticated recommendation systems, understanding deeper customer wants and needs remains challenging. Questions about whether customers are comparing costs or educating themselves on options are difficult to answer, complicating strategies for content and product display. This complexity affects sales, product development, and marketing, making it challenging to identify moments for impactful interventions.

Imagine a customer who is shopping for their groceries online from their regular grocer. She spends half an hour selecting 20 items and adding them to her basket, only to leave it all behind. This is curious behavior especially when the prices are competitive, products are available, shipment costs are low or zero, and there are no other clear deterrents in the customers' shopping experience.

Similarly, why does asking to open an account deter customers from completing their purchases? How does concern for the security of their credit card information affect their overall experience? What about factors such as being able to calculate the total order cost easily? How do all these factors affect their trust and sense of transparency affect abandonment rates? These were not mere queries but a narrative that played out repeatedly across many digital aisles.[1]

The complexity of human interactions and the social and economic networks that we are embedded in begs a deep understanding of the underlying factors that drive behavior and influence decisions. This is why asking the right questions is more important than ever, especially because of the availability of vast amounts of data with high ratios of noise to signal.

Game theory and behavioral economics, the disciplines I am trained in, offer powerful tools and frameworks to understand the complex dynamics of consumer behavior. These disciplines help us uncover hidden assumptions and ask the right questions to identify the true drivers of behavior amidst macroeconomic dynamics and the ever-changing landscape of consumer preferences.

Game theory, a branch of mathematics and economics, studies strategic interactions between individuals or groups where the outcome for each participant depends on the actions of all. It analyzes how decision-makers choose their actions to maximize their own benefits while considering the potential choices and reactions of others.

I believe every industry is a customer-centric industry and it is crucial to understand the motivations and drivers of the observed behaviors, whether they are internal stakeholders or external customers and consumers of products and services. It's not just about understanding isolated instances or scenarios; it's about discerning the intricate tapestry of interconnected use cases, and meticulously mapping and prioritizing the myriad questions that arise from them. Starting with why, through rigorous methodologies and a combination of descriptive, predictive, and causal analytics, we transition to crafting the most fitting models and solutions, guiding toward the optimal action, product, price, or service.

Simplifying the Complexity

At the end of my presentation, an executive, I'll call him John, representing a $1 billion consumer packaged goods (CPG) company approached me. In addition to overseeing a Data Center for Excellence, John also held a pivotal role as a senior executive in sales. He presented a perplexing scenario: During the pandemic, their strategic decision to elevate prices resulted in an unexpected surge in demand. The sales department, although proficient at demand forecasting, was now at a critical juncture. Given the unforeseen increase in demand post the price adjustments, they faced a consequential decision: Should they allocate a significant $100 million toward a new production facility, anticipating continued growth? Would consumer sentiment and purchase patterns remain strong as we approached the end of the pandemic? Or might there be a downturn, necessitating potential staff reductions?

Such dilemmas are not uncommon for professionals entrenched in the CPG sales and marketing sectors. They frequently grapple with substantial decisions, often relying on fragmented and partial data. This is not indicative of a reluctance to adopt a data-centric approach. Many are, in fact, keenly inclined toward evidence-based decision-making. The impediment often stems from data systems that, due to their fragmented structure, cannot seamlessly provide the needed insights. In many large enterprises, even generating a single report can be a prolonged process if it is not already automated. Compounding this challenge is the vast expanse of data, replete with occasionally conflicting insights that executives must meticulously navigate, much like the situation that John was dealing with.

The term *data-driven decision-making* is a buzzword in modern business, but its true meaning often gets lost. What does it really mean to be data-driven? Which insights are truly valuable for making sound decisions? The first steps in this process always go back to basics: identifying the core problem and asking the right questions.

In the vast sea of data, there are often just a few key points that, when connected, can solve even the most complex issues. It's not about having more data but about finding and understanding the most relevant pieces and how they fit together. This approach, focusing on quality over quantity, is what turns data into actionable insights.

The initial hurdle we confront is the identification, correlation, and combination of the right datasets. The act of curating the right data is intertwined with pinpointing the critical questions we seek answers to. It's akin to a cycle where the inception and conclusion blur into one another: Discerning the right questions invariably illuminates the path to identifying the right data, something that I have come to call "finding the minimum viable data (MVD)." MVD refers to those select data points, often subtle yet profoundly informative, which carry within them the crucial signals necessary for insightful analysis.

This cycle of asking the right questions and curating the right data points is fundamental. When done correctly, it leads to a more streamlined and efficient decision-making process. The key lies in understanding that more data is not always better. The real power comes from identifying those critical pieces of information that, when analyzed together, provide the most significant insights. This approach not only simplifies the data analysis process but also ensures that decisions are based on the most relevant and impactful information available. By focusing on MVD, businesses can cut through the noise, make more informed decisions, and ultimately drive better outcomes.

Reflecting on the conundrum of positive price elasticity provides a clear example of this principle. To truly understand the impact of rising prices on demand, we must first account for the correct control variables and influencing factors, such as seasonality, macroeconomic conditions, promotional prices, and merchandising. Once these elements are considered, we uncover the reality of negative price elasticity, aligning with established economic theory.

This example shows that merely having data on prices, shipments, or quantities sold is insufficient for gaining useful insights. Without the right data, we risk drawing wrong conclusions. This is crucial in an era when terabytes of data are generated daily, and many businesses struggle to make sense of it all. Often, they focus on cleaning and organizing data, creating numerous dashboards that end up causing more confusion than clarity. These dashboards fail to highlight the most important drivers of behavior or the factors that lead to sound decisions.

Some organizations run over 2,000 different reports simultaneously, often with conflicting insights. This clearly does not contribute to optimal decision-making. It's not surprising that some executives, overwhelmed by these dashboards and reports, end up relying on their gut instincts rather than validating their hypotheses with data. This underscores the importance of focusing on the MVD to guide decisions effectively.

Connecting the Dots

When the right datasets are connected and correlated, informed by the business hypothesis, a remarkable alignment emerges within and across departments. This was evident in John's paradox during our collaboration. Our first challenge was integrating the right datasets, both internal and external. They included units sold, competitor data, substitute products, and macroeconomic conditions. With a well-defined road map, we identified the essential datasets.

Once we had the data, we focused on identifying the right control factors to estimate price elasticity accurately. Contrary to initial beliefs, we confirmed that price elasticity was indeed negative. This analysis, powered by the MVD, enabled our client to analyze price elasticities for different retailers and products and strategize about cross-product and cross-retailer cannibalization effects.

This is when a more interesting question arose: Now that we have a handle on price elasticity, could we improve demand forecasting by including more information from marketing spend and product details? Understanding the demand for various products and SKUs across different grocers and regions helps better coordinate pricing, merchandising, and marketing. This highlights how one insightful question leads to another, forming the foundation of a cohesive and data-driven strategy.

By using ML algorithms, such as XGBoosts, as well as more sophisticated neural network models, we were able to significantly enhance the accuracy of our demand forecasting. In our instance, we saw an improvement from 75% accuracy to over 95%. This marked advancement can be attributed to various factors, including the detailed granularity of the data, a profound understanding of business nuances that informed the calculation of multiple features used as inputs for the algorithms, and a comprehensive knowledge of the ML algorithms themselves, enabling proper model design and tuning.

Improving demand forecasting by 20% for a multi-billion-dollar CPG company can yield significant financial, operational, and strategic benefits. Enhanced forecasting accuracy reduces carrying costs by optimizing inventory management, resulting in fewer stockouts and reduced excess inventory. This not only ensures consistent product availability, boosting sales and consumer loyalty, but also streamlines production and logistics, fostering operational efficiency. Financially, the company stands to increase revenue and improve profit margins through efficient inventory turnover and waste reduction. Furthermore, the ability to anticipate market demand allows for better strategic decision-making, fostering innovation and granting a competitive edge. Such advancements in forecasting can also strengthen supplier relationships, elevate retailer satisfaction, and bolster the company's sustainability efforts, making it a critical endeavor for any large CPG firm.

Uncovering Hidden Patterns: Models and Algorithms in Action

The path to uncovering hidden patterns and trends in data is paved with a variety of models and algorithms. These sophisticated tools enable us to transcend simple rule-based engines and dive deeper into the intricate relationships within the data.

Consider the advertisements you see on Instagram. They evolve over time, reflecting your changing interests as indicated by the reels you watch or the posts you read. This is a clear example of an ML algorithm at work, dynamically learning from data and adapting to provide relevant content. Contrast this with the static experience of opening a checking account online, where you might immediately receive a recommendation to open a savings account as well. This recommendation is driven by a rule-based engine, which operates on preset rules rather than adaptive learning.

Rule-based engines and ML algorithms serve different purposes in decision-making. A rule-based engine relies on explicit rules created by experts. These rules guide the system's operations, ensuring decisions align with predetermined logic. This approach offers transparency, because the reasoning behind each decision can be traced back to a specific rule. However, this clarity comes at the cost of flexibility. When the environment or data changes, the system does not adapt unless the rules are manually updated, which can be time-consuming and labor-intensive.

By contrast, ML algorithms "teach" systems to recognize patterns in data. Rather than being programmed with explicit instructions, these algorithms derive rules from large datasets. Once trained, they can predict outcomes for new, unseen data. Their adaptability is their greatest strength; they can be retrained with new data to accommodate changing conditions. However, this flexibility often comes with a trade-off in transparency. Especially with complex models like deep neural networks, understanding why a particular decision was made can be challenging, making these systems appear as "black boxes."

In essence, although rule-based engines are ideal for well-defined, logical operations requiring clear traceability, ML algorithms excel in dynamic environments where data patterns are complex and constantly evolving. The choice between them depends on the task at hand, the level of transparency required, and the nature and complexity of the data.

Understanding these nuances is crucial for executives navigating the data-driven landscape. Leveraging the right model can uncover profound insights, drive more informed decision-making, and ultimately create significant value. As we delve into various applications of AI and ML, it's essential to recognize the strengths and limitations of each approach, ensuring we harness their full potential to address our unique business challenges.

Decoding Consumer Behavior: The Interplay of Psychology and Economics

This is only half the story. How many individuals have you encountered who have opened a savings account prompted by a pop-up after initiating a checking account? Both rule-based engines and traditional ML algorithms often falter in influencing consumer behavior because they fail to account for the diverse motivations, reasons, and incentives driving different customers. Consumer reactions vary significantly based on their experiences. For instance, consider the advertisement they encountered just 30 minutes prior to considering opening a bank account. If they had been enticed by a shoe or handbag advertisement on Instagram, they are likely not in the mood to open a savings account!

Human decision-making is multifaceted, influenced by emotional, psychological, social, and contextual factors. A set of pre-defined rules or even advanced ML models will not capture this complexity comprehensively. For a strategy to be effective, it often needs to be tailored to the individual. Rule-based systems, especially if not intricately designed, might apply a one-size-fits-all approach, which can be less effective than personalized strategies derived from data-driven insights. Similarly, traditional ML models, while adaptive, can still fall short in understanding the deeper, often irrational human behaviors driving decisions.

These algorithms operate within certain confines. At the heart of this challenge is the multidimensional nature of human decision-making. Decisions are rarely purely rational; they are interwoven with emotions, cultural norms, past experiences, and even seemingly illogical biases. The static structure of rule-based systems struggles to keep pace with the dynamism of consumer preferences that evolve with societal trends, technological advancements, and shifts in cultural values. ML models, although they learn from data, can sometimes miss the nuances of human behavior, especially if the data fed into them lacks depth or context.

For instance, a model might suggest offering discounts to customers who have not purchased in a while. But without deeper insights into why that customer has not purchased, the tactic might miss the mark. Additionally, the transparency of rule-based systems, though a strength in many contexts, can backfire when consumers feel they are being manipulated by a predictable set of rules, leading to decreased trust. ML models, with their black-box nature, can also erode trust if users do not understand how decisions are made.

Modern insights into consumer behavior lean heavily on the analysis of vast, often unstructured datasets, a realm where both rule-based engines and traditional ML algorithms without analytical augmentations falter. By leveraging principles from behavioral economics, such as understanding the role of loss aversion (where people prioritize avoiding losses over acquiring gains) or the anchoring effect[2] (where initial information influences subsequent decisions), strategies can be more finely tuned to align more closely with real-world decision-making patterns, nudging consumers in desired directions while respecting the intricacies of human choice.

Imagine a chocolate company that introduces a new product at a higher price point, establishing a perception of it being a luxury item. After some time, they could offer a "discount" that still brings in a healthy profit margin. Consumers, anchored to the initial higher price, might perceive this as a great deal, increasing their likelihood to purchase. This is the anchoring effect in practice. Of course, the trick is still in identifying the price and the discount that will have the desired sales volume.

Think about another example: A product's packaging or promotional materials can feature testimonials such as quote from a famous author on a book or "influencers" on Instagram or other social media channels, ratings like "Heather's Choice" in Indigo bookstores, or even simple statements like "best-selling flavor" to leverage the power of the crowd and sway potential buyers. This is called *social proof*: People often look to the actions of others to guide their own decisions, especially in uncertain situations.

Loss aversion is another very interesting concept that is well studied by behavioral economists: People feel the pain of losing something more intensely than the pleasure of gaining something of equal value. An application of this concept in limited-time offers or limited edition flavors can be introduced. By emphasizing that consumers might "miss out" if they do not act quickly, the brands tap into the fear of loss, driving urgency and increasing sales.

These examples illustrate how behavioral economic principles can be leveraged. These principles provide insights into the often irrational ways in which consumers make decisions. By understanding and tapping into these inherent psychological biases and tendencies, companies can design more effective strategies to influence consumer behavior.

Empowering Behavioral Economics: The Synergy of Data Analytics, ML, and AI

Behavioral economic principles are indeed powerful especially in understanding and influencing the often-irrational nuances of consumer decision-making processes. However, understanding the effectiveness of strategies rooted in behavioral economics poses several challenges. One main challenge arises because of the complexity of human behavior: Even if behavioral economic principles have been employed, human behavior remains multi-faceted and influenced by numerous external factors.

Disentangling the impact of a specific strategy from other influencing factors requires sophisticated data analytics. If not properly controlled, confounding variables can cloud the true effectiveness of a strategy. However, different strategies might work differently across various segments of the population.

What's effective for one group might not be for another. This necessitates collecting and analyzing data at both a granular and aggregate level.

Temporal dynamics also matter and make the situation even more complicated: Consumer preferences and behaviors evolve over time, and as a result, a strategy's effectiveness might wane or evolve over time. This makes it vital to continuously collect and analyze data, tracking shifts over time. Seasonal patterns, trend cycles, or even sudden events can alter consumer behavior, requiring adaptive strategies.

The COVID-19 pandemic serves as a poignant example of a major event that significantly affected the temporal dynamics and preferences of consumers. As the pandemic initially unfolded, there was a sudden surge in demand for essential products, such as hand sanitizers, masks, and nonperishable food items.[3] Consumers also quickly adopted or increased online shopping, leading to a surge in e-commerce. As lockdowns and work-from-home became prolonged, there was a noticeable uptick in the demand for home office equipment, home fitness gear, and digital entertainment services. Trends like do-it-yourself home improvement or baking also gained traction. Even as many regions started recovering, some changes seemed to have a lasting impact. For example, the adoption of e-commerce, telehealth services, and remote work tools might have set new baselines for certain industries.[4]

Concurrently, consumer preferences underwent significant evolution. Post-pandemic, many consumers exhibited a willingness to pay a premium for safer shopping experiences. Retailers offering contactless deliveries, curbside pickups, or enhanced in-store safety measures found favor. Additionally, with global supply chain disruptions and a desire to support local businesses facing economic challenges, there was a marked shift toward local purchasing. Digital adoption was no longer a luxury but a necessity. From elderly individuals learning to shop groceries online to businesses adopting digital tools for remote collaboration, digital transformation affected nearly every demographic and industry. However, the economic downturn made consumers more conscious of their spending. There was a noticeable shift toward value-driven purchases, with consumers seeking products and services offering durability, sustainability, and long-term value. The pandemic's onset and its subsequent waves influenced both immediate reactive behaviors and longer-term adaptive behaviors.

These examples illustrate that understanding and harnessing the effectiveness of behavioral economics strategies is a nuanced endeavor. The marriage of these strategies with robust data analytics, ML, and AI can offer potent solutions and deeper insights.

By analyzing vast datasets, we can gain insights into multifaceted consumer behavior patterns. Descriptive and inferential statistics can reveal underlying tendencies and correlations. Additionally, methods like time-series analysis can reveal deep insights into evolving consumer behaviors, enabling businesses to discern underlying patterns and adapt to changing trends. Furthermore, A/B testing can empirically evaluate the impact of different strategies on consumer engagement, guiding more informed decisions.

ML elevates this understanding to a predictive plane. Techniques like unsupervised learning can segment consumers, enabling personalized strategies and identifying niche consumer groups and their specific preferences. Additionally, recommendation systems, like those used by Netflix and Amazon, can suggest products or services tailored to individual consumers based on their past behavior, exemplifying the power of personalization. I will dive into these models and techniques in the next few chapters.

Moreover, the challenge of validating the real-world effectiveness of behavioral strategies can be addressed through regression models that can quantify the impact of specific strategies on sales, user engagement, or other key performance indicators, controlling for other influencing factors. Adaptive experimentation, powered by AI, can further optimize testing by dynamically adjusting parameters to rapidly hone in on the most effective strategies.

Natural language processing can analyze consumer sentiments from textual data like reviews or feedback, providing deeper insights into emotional and contextual factors influencing behavior. Additionally, advanced AI models can dynamically adapt marketing content, promotions, or user interfaces in real time based on individual user behavior.

From dynamically adapting content in real time based on individual user behavior using reinforcement learning algorithms to ensuring that strategies are ethically sound and free from biases, AI bridges theory and practice. In essence, behavioral economics lays the theoretical groundwork, and the trio of data analytics, ML, and AI translates these insights into actionable, adaptable, and effective real-world strategies.

Crafting a Customer-Centric Paradigm: The Fusion of Technology and Behavioral Insights

Combining best practices in data analytics, ML, AI, and behavioral economics creates a holistic approach that addresses both the quantitative and qualitative aspects of customer behavior, making it a potent strategy for customer-centric industries.

Data analytics provides a foundational understanding of customer behaviors by examining past trends, segmentations, and patterns. This empirical backbone ensures businesses have a grounded view of their customer base. ML then builds on this by predicting future behaviors, enabling businesses to anticipate and proactively cater to customer needs. Personalization, a vital aspect of customer-centric models, is significantly enhanced by ML algorithms, enabling businesses to tailor offerings to individual customer preferences.

AI streamlines and synergizes these processes, offering dynamic adaptability and real-time responses. Its capability to process vast amounts of data rapidly ensures that insights are timely and relevant.

Behavioral economics complements these technological tools by introducing a deep understanding of the often-irrational human decision-making processes. While the tech tools process the *how* and *what* of customer behavior, behavioral economics sheds light on the *why*. By integrating insights about cognitive biases, emotions, and social factors, businesses can craft strategies that resonate more deeply with human psychology.

The convergence of data analytics, ML, AI, and behavioral economics offers a comprehensive approach to understanding and influencing customer behavior. For customer-centric industries, this union ensures they are not only meeting customers' needs based on data but also connecting with them at a deeper, more human level, ensuring long-term loyalty and engagement.

CHAPTER 2

It Is All Connected
Behavioral Economics, Decision-Making, Biases, and Heuristics

"We can be blind to the obvious, and we are also blind to our blindness."

– Daniel Kahneman

I met Rob on LinkedIn in May 2020. Rob was the chief customer experience (CX) and innovation officer at one of the largest credit unions in Canada, and I found myself intrigued both by his title and his background in behavioral economics. When Rob and I connected, one of the first things that he shared with me was that his team "had all the data, but did not really know their members well enough to drive their financial well-being." This struck a chord with me, because just a few months earlier, I had heard the chief CX officer at one of the five largest banks in the United States echo the exact same sentiment. I wondered why, or what it even means to a bank to "know" a member or a customer well enough to drive their financial well-being. In other words, what aspects of a member's behaviors indicate their needs, desires, preferences, or the risks that they might be exposed to, and just as importantly, how to extract signals for these attributes from the huge volume of data that a bank has access to.

History and Origins of Behavioral Economics

These questions are at the heart of behavioral economics: a field of study that explores economic decision-making through the lens of human psychology. Unlike traditional economics, which often assumes that individuals are rational agents who make decisions to maximize utility, behavioral economics acknowledges that people act according to various cognitive biases and emotional factors – not just pure logic. In fact, behavioral economics emerged as a response to the limitations of classical and neoclassical economic theories that are rooted in the assumption of pure rationality, treating individuals as rational utility-maximizing "agents" rather than human beings.

Richard Thaler, who was awarded the Nobel Memorial Prize in Economic Sciences for his contributions to behavioral economics in 2017, refers to these economic agents as *Homo Economicus* or *Econs* for short.[1] He uses this term to describe the theoretical concept of a perfectly rational and fully informed agent who optimizes utility or profit in every decision. Thaler contrasts this idealized agent with real humans, whom he often refers to as *Humans*, to highlight the discrepancies between the theoretical models of classical economics and actual human behavior.

Early Days

The history of behavioral economics can be traced back to various intellectual developments that occurred long before the field was formally recognized. These early foundations laid the groundwork for understanding economic decision-making in a way that incorporates psychological factors, challenging the traditional economic assumption of fully rational actors.

These foundations were laid by a diverse group of thinkers who introduced key concepts highlighting the role of psychological and social factors in economic decision-making.

One of the early contributors was Thorstein Veblen, an American economist and sociologist, who introduced the idea of "conspicuous consumption" in his 1899 book *The Theory of the Leisure Class*.[2] This concept suggests that people's spending is often driven more by social status than by utility maximization. Think about people you know who have spent $100 000 on a watch! More on that and "signaling behavior" later.

Similarly, British economist John Maynard Keynes, in his 1936 work *The General Theory of Employment, Interest, and Money*,[3] introduced the concept of "animal spirits" to explain how psychological factors influence economic behavior, especially under uncertainty.

A significant leap in this direction was made by Herbert Simon, an American economist and political scientist, who introduced the concept of "bounded rationality" in the 1950s.[4] Simon argued that human decision-making is limited by available information, cognitive limitations, and time constraints, challenging the notion of perfect rationality. His work laid the foundation for more realistic models of human behavior in economics. Additionally, Hungarian American psychologist George Katona's work in economic psychology contributed to understanding the psychological aspects of consumer behavior, further blending economics with psychology.[5] These early developments were crucial in highlighting the limitations of traditional economic theories and set the stage for the emergence of behavioral economics as a distinct and influential field in the latter part of the twentieth century.

During the 1960s and 1970s, the field of behavioral economics began to take a more defined shape, largely due to the groundbreaking work of psychologists Daniel Kahneman and Amos Tversky. Their research focused on heuristics and biases, challenging the traditional economic view that people are rational actors who make decisions to maximize their utility. They showed that people often rely on cognitive shortcuts (heuristics) that lead to systematic errors in judgment and decision-making.[6]

Kahneman and Tversky identified several key heuristics, such as the "availability heuristic," where people assess the probability of events based on how easily examples come to mind, often leading to overestimation of frequent or recent events. Another notable heuristic is the "representativeness heuristic," where people judge the likelihood of an event by how closely it resembles their existing stereotypes, disregarding actual statistical probability. Their work also introduced the concept of "loss aversion" in *Prospect Theory* in a 1979 paper,[7] which posits that people tend to prefer avoiding losses over acquiring equivalent gains, demonstrating that the pain of losing is psychologically more impactful than the pleasure of gaining. They also explored how the way information is presented (or "framed") can significantly affect decisions and judgments. People react differently to a choice depending on whether it is presented as a loss or a gain.

This body of work had a profound impact across multiple disciplines, extending far beyond economics to psychology, law, medicine, and political science. It led to a deeper understanding of human behavior, emphasizing the role of psychological factors in economic decision-making. Kahneman's contributions, in particular, were recognized with the Nobel Memorial Prize in Economic Sciences in 2002. Their research fundamentally altered how economists and policymakers understand and predict human behavior in economic contexts, highlighting the importance of psychological insights in the development of economic theory.

Entering Mainstream Economics

The integration of behavioral economics into mainstream economics, significantly propelled by the works of Daniel Kahneman and his coauthors, marked a pivotal shift in economic thought. Initially, the field of economics was dominated by models that presumed individuals as rational actors, consistently making decisions that maximize their utility. However, the empirical findings and theoretical frameworks developed by Kahneman, Tversky, and others challenged this paradigm by demonstrating systematic deviations from rationality due to cognitive biases and heuristics.

Kahneman and Tversky's seminal contributions, such as prospect theory and studies on loss aversion and the framing effect,[8] provided robust evidence that human decisions often defy traditional economic predictions. These insights were initially met with skepticism but gradually gained acceptance as their applicability across various domains – ranging from finance and marketing to public policy and health – became evident. The behavioral approach advocated for a more nuanced understanding of economic decisions, emphasizing the importance of psychological factors and real-world complexities.

The formal recognition of behavioral economics within the mainstream came in part through institutional developments, such as the establishment of the *Journal of Behavioral Economics* in 1976 (later renamed the *Journal of Behavioral and Experimental*

Economics), and the awarding of the Nobel Memorial Prize in Economic Sciences to Kahneman in 2002, and later to Richard Thaler in 2017, both for their contributions to behavioral economics. These recognitions signaled a growing presence of behavioral insights in economic research and policymaking.

Thaler's work has spanned several key areas, including savings behavior, choice architecture, and the concept of "nudging,"[9] among others. His Econs versus Humans concepts have significantly shaped the understanding of human behavior in economics, highlighting the impact of cognitive biases and social preferences on decision-making. Econs are depicted as always making rational choices, unaffected by biases or emotions, and possessing complete information. In reality, Thaler's work in behavioral economics shows that real people (Humans) often deviate from this ideal due to various cognitive biases, limited self-control, and other psychological factors. This contrast is a central theme in Thaler's work, emphasizing the importance of considering real human behavior in economic theory and practice.

The application of behavioral economics to public policy, particularly through the concept of nudges as also popularized by Thaler and Sunstein in their book *Nudge*, has shown how small changes in the choice architecture can have significant impacts on behavior without restricting freedom of choice.[10] Examples of nudges include the placement of healthier foods at eye level to encourage better eating habits or default enrollment in pension plans to increase retirement savings.

This approach has been adopted in various governmental and nongovernmental policy initiatives globally. Notably, David Cameron's government in the United Kingdom established a behavioral insights team in 2010 that became widely known as the "Nudge unit," with the aim of applying behavioral economics principles to design better policies and services. This unit played a large role, along with health sectors, in crafting government responses to the COVID-19 pandemic.[11]

Another notable contribution by Thaler and his coauthors is the development of the theory of mental accounting.[12] This concept describes how people categorize and treat money differently based on subjective criteria, such as the source of the money or the intended use, which can lead to irrational financial behaviors. He also codeveloped the "Save More Tomorrow"™(SMarT) program with Shlomo Benartzi, which helps employees increase their savings rates over time by committing in advance to allocate a portion of their future salary increases toward retirement savings. This program leverages behavioral insights to overcome procrastination and loss aversion.[13]

Thaler has also explored the implications of behavioral economics for financial markets, challenging the efficient market hypothesis by showing how anomalies and irregularities in market behavior, investor behavior, and financial crises can be explained by psychological factors. His work in this area challenges the efficient market hypothesis[14] by showing how psychological factors can lead to predictable patterns in financial markets and has shed light on phenomena such as the equity premium puzzle and the behavior of stock market prices in ways that traditional economic models cannot fully explain.

The journey of behavioral economics from the fringes to the mainstream of economic thought underscores the complexity of human behavior and the need for economic models that reflect this reality. Kahneman, Tversky, Thaler, and their contemporaries not only expanded the toolkit of economics but also bridged the gap between economics and psychology, leading to richer, more nuanced understandings of how economic decisions are made.

Subsequently, the field has continued to grow, influencing not only academic research but also practical policymaking, with behavioral insights units being established in governments around the world to apply these principles to improve public policies and services.

Current Research and Practical Applications

The landscape of behavioral economics research in academia is vibrant and continuously evolving, with scholars exploring a wide array of topics that intersect with daily life and global challenges. One notable area of investigation is the realm of digital and platform economics, where researchers delve into the intricacies of online behavior. With the rise of digital platforms, researchers are examining how behavioral economics principles apply to online behaviors, including e-commerce decision-making, social media interactions, and digital content consumption. Studies[15] explore how digital nudges can influence online shopping habits, privacy concerns, and the dissemination of information or misinformation. A specific example of this is the study of digital nudges in e-commerce platforms to understand how varying the presentation of product recommendations influences consumer purchase decisions.[16]

Dan Ariely, another leading behavioral economist with an extensive research background in consumer behavior, delves into the psychological underpinnings of how consumers make purchasing decisions, with a particular focus on how elements like pricing strategies, product presentation, and the checkout process can significantly influence these choices. His work in this area leverages principles from behavioral economics to explain why consumers behave in ways that might seem irrational from a traditional economic perspective.

Ariely has explored how different pricing strategies affect consumer perception of value and willingness to purchase. In his 2008 book *Predictably Irrational: The Hidden Forces That Shape*

Our Decisions,[17] Ariely details a notable experiment showing how arbitrary pricing ("anchoring effect") can influence how much consumers are willing to pay for items. By presenting consumers with a price that serves as an anchor, their subsequent valuation of products is influenced, even if the initial price is unrelated to the product's actual value. This research has implications for how companies set prices for new products, suggesting that initial prices can have a long-lasting impact on perceived value.[18]

His research also examines how the way products are presented to consumers can affect their choice. This includes the physical layout of products in a store or on a website, the assortment of products offered, and how information about the products is provided. For example, he has shown[19] that offering too many choices can lead to "choice overload," reducing consumers' satisfaction with their decision and sometimes leading them to make no purchase at all. This has led companies to rethink strategies about product variety and how choices are presented to consumers.

There's also a growing body of research focused on applying behavioral economics to health care and tackling pressing public health issues. This includes studies on how to increase the uptake of vaccinations,[20] improve adherence to medical treatments,[21] and encourage healthy lifestyle choices.[22] Researchers are using nudges and other behavioral interventions to address challenges such as obesity, smoking, and drug adherence.[23] For instance, research has demonstrated the effectiveness of text message reminders in increasing flu vaccination rates among the population,[24] showcasing a simple yet powerful application of behavioral economics in enhancing public health outcomes.

Dan Ariely's contributions to health care research through the lens of behavioral economics have focused on enhancing patient behaviors, treatment adherence, and health outcomes.[25] His work encompasses a range of critical areas, including medication adherence, pain management, health behavior change, patient decision-making, and critiques of the health care system itself.[26]

His studies[27] on medication adherence reveal the psychological barriers preventing patients from following prescribed treatments, despite understanding their benefits. He proposes interventions like simplifying medication schedules and employing reminder systems, addressing the cognitive biases leading to non-adherence. In pain management, Ariely explores how expectations and prior experiences can influence pain perception, suggesting that health care providers' communication strategies can significantly affect pain management outcomes.[28]

Furthermore, Ariely has delved into motivating healthier lifestyles by applying behavioral economic principles such as immediate rewards and social proof strategies. His research includes the use of commitment devices[29] to encourage and sustain healthy behaviors over time. In examining patient decision-making, Ariely's work sheds light on how patients value treatment options and how their choices are influenced by emotions like fear and hope, advocating for enhanced shared decision-making processes between patients and health care providers.[30]

Cross-cultural behavioral economics is another expanding field, examining how cultural differences affect economic decision-making. Research comparing saving behaviors across cultures has unveiled substantial variations, attributing these differences to societal norms and values that influence financial planning and risk preferences.[31]

Financial decision-making, particularly in the context of emerging technologies like cryptocurrencies, is also a hotbed of behavioral research. Devkant Kala's 2023 study[32] investigated the psychological factors driving young people's investments in cryptocurrencies, revealing that the fear of missing out plays a significant role in individuals' decisions to invest in these volatile digital assets.

It Is All Connected

Neuroeconomics stands at the exciting intersection of economics, psychology, and neuroscience, seeking to uncover the neural underpinnings of economic decisions. Richard L. Peterson's 2005 study[33] used functional magnetic resonance imaging (fMRI) to observe brain activity during investment decisions, identifying specific brain regions that are activated when individuals face financial risks.

Ariely's research on financial decision-making, particularly regarding the "pain of paying" and its implications for savings and spending habits, delves into the psychological aspects of how individuals interact with money.[34] This area of study is a vital component of behavioral economics, because it provides insights into why people can make financial decisions that do not align with their long-term best interests.

The concept of the pain of paying refers to the psychological discomfort or pain people experience when they part with their money. Ariely's research[35] suggests that this pain varies depending on the mode of payment and the context of the transaction. For instance, paying with cash typically induces more pain than using credit cards because cash transactions are more tangible and immediate, making the loss felt more acutely. This tangibility and immediacy of cash payments can actually lead to more restrained spending behaviors.

Beyond the implications for spending, Ariely's research also extends to saving behaviors. By applying behavioral economics principles, he has explored mechanisms to make saving feel more rewarding and less like a sacrifice. One approach has been to leverage instant gratification tendencies by providing immediate rewards or feedback for saving money, thus counterbalancing the pain of paying or the reluctance to save.

Ariely's collaborations with financial institutions have involved applying these and other insights to real-world settings, designing interventions that can nudge individuals toward more responsible financial behaviors. For instance, altering the presentation of saving options to make the future benefits more tangible and immediate can encourage higher saving rates. Similarly, financial products that require more active decision-making to spend money (thereby increasing the pain of paying) can help reduce unnecessary expenditures.

Back to the Beginning

When I met Rob on LinkedIn, and later the co-CEOs of the credit union, the same questions kept coming up, such as

> *"We have access to a vast amount of data, but how do we use it to improve the financial well-being of our members?" "How can we use it to understand what they really need and where in their journey they are?" "How can we encourage them to save more using vehicles such as term deposits and other saving accounts?" "How can we offer loans at tailored interest rates to make sure that our members are engaged with the right products?" These are indeed difficult questions and have been at the core of the research agenda for many behavioral economists and social scientists. Although these difficulties are found in huge masses of data, that same data holds all the solutions – provided you have the tools to sift through it.*

Let me walk you through a thought exercise to demonstrate the complexity and also the opportunity of the situation.

A typical individual usually banks with two or three different institutions, and their banking behavior can vary widely based on their personal needs, financial goals, and available options. An individual often has a primary bank where they maintain their checking and savings accounts. This bank is usually chosen for its convenience, like having nearby branches or good online banking services. They use their checking account for day-to-day transactions like receiving a salary, paying bills, and making

It Is All Connected

purchases. The savings account is used to set aside money for future needs, emergencies, or specific goals, often earning a small amount of interest.

However, when buying a house, individuals often shop around for the best mortgage rates and terms. They might end up choosing a different financial institution for their mortgage, one that offers lower interest rates or better loan terms than their primary bank.

For student loans, individuals often go with federal or private lenders who specialize in educational loans and offer competitive rates. Car loans might be obtained through auto finance companies, credit unions, or banks. The choice often depends on who offers the best interest rates and terms at the time of the car purchase.

Additionally, individuals often have credit cards, which might or might not be from their primary banking institution. Credit card choices are influenced by factors like interest rates, reward programs and additional benefits, and annual fees. With the rise of fintech, many Gen Z customers also use online-only financial services for specific needs like peer-to-peer payments, budgeting apps, or high-yield savings accounts.

For investment purposes, individuals might have accounts with brokerage firms or investment banks different from their primary bank. Retirement accounts like IRAs or 401(k)s might be managed through their employer or an independent financial services company. Furthermore, insurance products, like life or auto insurance, are usually obtained from insurance companies rather than banks. Some individuals also explore other financial products like annuities or personal lines of credit from various financial institutions.

The initial hurdle in the banking sector is that, despite having extensive information about their clients, financial institutions, including primary banks, possess only a fragmented insight into their customers' complete needs and preferences. This situation is often described in economic terms as a scenario of incomplete information, characterized by inherent biases within the available data. A bank's understanding is limited to the observable interactions: the products and services their customers actively engage with and prefer. However, this does not encompass a comprehensive view of potential needs that the bank's current offerings might not meet. The data also fails to illuminate areas of ignorance among customers regarding existing products or services. In essence, a bank might have the ideal solution for a customer's needs and yet remain blind to the customer's lack of knowledge about these available options. This creates a strong case for implementing strategies like product recommendation and behavioral segmentation, more of which I'll discuss later.

Additionally, the data that banks possess is restricted to observable behaviors, providing a limited perspective on customer interactions. They capture only a subset of customer behavior: the specific products that customers engage with at that particular institution. This limitation echoes the bias observed in survey data,[36] where responses are often skewed toward extreme experiences – customers are more likely to report extreme dissatisfaction or satisfaction. Such responses result in self-selection bias, because they do not accurately represent the full spectrum of customer experiences or needs. This narrow dataset overlooks the multifaceted and evolving financial requirements of customers, which extend beyond the confines of their interactions with a single financial institution.

All of these limitations in the data, and hence the insights about their customers that a financial institution experiences, are due to the choices that individuals make. The reasons behind these decisions and the psychology of decision-making will shed light on the interventions, engagement, and product strategies that will help banks and financial institutions interact with their customers at the right moment with the right products and services.

Despite having access to extensive data, banks often find it disconnected, lacking a comprehensive view of customer behaviors across different financial products and services.

Psychology of Decision-Making

The psychology of decision-making is a rich field that explores how people make choices, the processes underlying these choices, and the factors that influence decision outcomes. This interdisciplinary area draws from psychology, economics, neuroscience, and sociology to understand the complexities of human behavior.

Dual-Process Theories

One of the foundational frameworks in decision-making psychology is the concept of dual-process theories. These theories propose that humans use two distinct systems for processing information:

System 1 is fast, automatic, intuitive, and emotional; it operates effortlessly and without deliberate control. This system handles everyday decisions, relies on heuristics (mental shortcuts), and can be influenced by biases. It's responsible for the quick judgments and reactions we have to the world around us, often without our conscious awareness.

System 2 is slow, effortful, logical, and calculating. It requires conscious effort and is engaged when we need to solve complex problems, make judgments that require attention, or when we evaluate choices more critically. System 2 is what we typically think of as "rational" decision-making, but it's resource-intensive and can be lazy, often deferring to System 1 unless specifically engaged.

These concepts were popularized by Daniel Kahneman in his seminal book *Thinking, Fast and Slow*,[37] which details how these two systems interact to shape our judgments and decisions, often leading to systematic errors and biases.

When you see a face and instantly recognize the person or perceive their emotions (e.g. happiness or anger), it's System 1 at work. This process is automatic and does not require conscious thought. For an experienced driver, driving on a familiar road or performing routine actions in the car happens almost automatically, with little need for active concentration. First impressions, such as deciding whether a person you have just met seems trustworthy or friendly, is also a function of System 1. Understanding sentences or recognizing words in your native language also occurs effortlessly and without deliberate analysis, and, no matter how bad at math you are, the answers to basic problems like 2 + 2 come to mind immediately and without effort. These are all examples of System 1 at work.

However, solving a challenging math problem that requires you to stop and think, perhaps even grabbing a piece of paper to work through it, engages System 2. When you are trying to learn something new, such as playing a musical instrument or acquiring a new language, System 2 is heavily involved in deliberate practice and attention to detail. This is exercised when evaluating the pros and cons of a significant decision, like buying a home or making a career change, which requires analytical and reflective capabilities. Engaging in strategic games like chess, especially when planning several moves ahead, relies also on the slow, deliberate thought processes characteristic of System 2.

Understanding the interplay between these systems can help us recognize when we might be relying too heavily on intuition or when a more thoughtful analysis is warranted.

Let us say you are buying a new car and find yourself strongly drawn to a particular model because it's the same make and color as your first car, which you loved. Your initial, emotional reaction is a classic System 1 response, driven by nostalgia and a positive association with the color and brand. However, recognizing the interplay between System 1 and System 2, you decide to engage System 2 to ensure you are making a sound decision. You start researching the car's reliability, fuel efficiency, customer reviews, and safety ratings, and compare it with other models. This process might reveal that although your initial choice was emotionally satisfying, another car offers better value and performance for your needs.

In this scenario, understanding the interplay between System 1 and System 2 enables you to recognize when an intuitive choice needs to be balanced with more analytical thinking. It helps in preventing decisions based solely on gut feelings or biases, ensuring a more thorough evaluation of options. This balanced approach to decision-making leverages the rapid, intuitive responses of System 1 when appropriate, while also harnessing the analytical power of System 2 to make well-informed choices, particularly in complex or high-stakes situations.

Heuristics and Biases

Heuristics are mental shortcuts that simplify decision-making under uncertainty but can lead to systematic biases. Kahneman and Tversky's seminal work identified several key heuristics, such as the availability heuristic, the representativeness heuristic, and the anchoring heuristic. Heuristics and biases are critical concepts in the study of logic. These shortcuts can significantly affect the rationality of human decision-making, and their identification helps to explain systematic deviations from logic or rational choice theory.

Heuristics are simple, efficient rules, either hardwired or learned, used to form judgments and make decisions. They reduce the cognitive burden associated with decision-making and can be incredibly helpful with quick decision-making, as well as leading to errors or biases.

For example, the availability heuristic leads individuals to judge the likelihood of events based on how easily examples come to mind. If someone frequently hears news stories about airplane accidents, they might overestimate the risk of flying, because these vivid and recent incidents are easily recalled, making them seem more common than they actually are. Similarly, the representativeness heuristic involves making judgments about the probability of an event by comparing it to an existing mental prototype. For instance, on meeting someone who is shy and enjoys reading, people might quickly assume this person is a librarian rather than a salesperson, ignoring statistical realities about the prevalence of these professions. This heuristic focuses on the similarity to a category rather than using logical statistical assessment. Another example is the anchoring heuristic, when individuals rely heavily on the first piece of information they receive. For instance, if the first car a person looks at is priced at $30 000, this figure sets an anchor, and all subsequent cars are judged relative to this price, regardless of their actual value or features. This initial anchor can skew perception and decision-making, often leading to irrational financial decisions. Last, the simulation heuristic depends on our ability to mentally simulate an event; the easier it is to imagine, the more likely we believe it to happen, influencing our expectations and risk assessments. These heuristics illustrate the mental mechanisms that underpin our judgments and decisions, highlighting the delicate balance between cognitive efficiency and irrationality.

Biases are systematic patterns of deviation from norm or rationality in judgment. They arise as a result of cognitive processes that simplify information processing. Biases often stem from heuristics but can also occur in other ways.

For instance, confirmation bias reflects our tendency to favor information that confirms our existing beliefs or hypotheses, while neglecting or undervaluing evidence that contradicts them. For example, when researching a potential investment, an investor might give more weight to positive reviews and optimistic projections that support their decision to invest, overlooking critical reports and negative indicators. This bias can lead to poorly informed decisions based on incomplete information. Overconfidence bias is another pervasive issue, occurring when individuals overestimate their knowledge, ability, or access to information. An example of this can be seen in the realm of financial trading, when traders might believe they can predict market movements better than the average investor, often leading to risky trades that do not pay off. This overestimation of one's own decision-making capabilities can result in significant financial losses and missed opportunities for more prudent investment strategies.

Loss aversion, a concept rooted in prospect theory, illustrates our tendency to prefer avoiding losses rather than acquiring equivalent gains. For example, a person might refuse to sell an underperforming stock at a loss, even if rational analysis suggests that selling it would be the best decision. The pain associated with realizing a loss outweighs the logical assessment of the situation, leading to irrational holding patterns that can exacerbate financial harm.

These biases in decision-making often lead us down paths that deviate from rational judgment, influenced by a variety of psychological factors. The status quo bias, for instance, reveals our preference for maintaining current conditions or sticking with decisions previously made, even when change might be beneficial. This is closely related to the overconfidence bias, when individuals tend to overestimate the accuracy of their beliefs and judgments, confidently navigating decisions without due consideration of their actual knowledge or the uncertainty involved. Another intriguing bias is hindsight bias, the tendency to view past events as having been more predictable than they truly were, leading us to believe we knew the outcomes all along. Last, the Dunning-Kruger effect is a phenomenon in which individuals with limited knowledge or ability in a specific domain overestimate their own skills, not recognizing their deficiencies.

Additionally, the endowment effect highlights how ownership increases an item's perceived value; simply owning something can make it seem more valuable than if we did not own it. Then there's the sunk cost fallacy, which drives us to continue investing in a project or decision purely because of the resources already committed, regardless of the current or future benefits. The framing effect demonstrates how the same information can lead to different conclusions based on how it's presented, affecting our choices and judgments. Additionally, the halo effect illustrates how our overall impression of a person can be influenced by judgments in unrelated areas, such as assuming someone is inherently good in many aspects because they excel in one. Together, these biases underscore the complex psychology behind our decision-making processes, often leading us away from logical conclusions in favor of judgments shaped by these varied and sometimes misleading influences.[38]

Recognizing heuristics and biases in our decision-making process is crucial because it enables us to identify when we might be relying too heavily on intuition or when a more analytical approach is warranted. For example, if you are making a significant financial decision and find yourself leaning strongly toward an option because it's the most readily available in your mind (availability heuristic), it might be a signal to step back and engage in a more thorough analysis (System 2 thinking). Similarly, if you are evaluating evidence that supports your preexisting beliefs while dismissing evidence to the contrary (confirmation bias), acknowledging this bias can prompt you to seek out and consider information more impartially.

Understanding these concepts not only enhances personal decision-making but also has important implications for fields ranging from economics and finance to public policy and health care, where recognizing and mitigating biases can lead to better outcomes.

Noise

In his more recent work *Noise: A Flaw in Human Judgment*, Kahneman (along with coauthors Cass R. Sunstein and Olivier Sibony) expands beyond the dual process theory to explore the concept of "noise" in decision-making. Although much of Kahneman's earlier work focused on biases – systematic errors that affect human judgment in predictable ways – *Noise* focuses on the variability of judgments that should ideally be identical. Kahneman and his coauthors differentiate between bias (a directional deviation) and noise (random scatter) in judgments, arguing that both can lead to errors in decision-making, but organizations and individuals have largely overlooked the impact of noise.

"Noise" refers to the unwanted variability in decisions that should be uniform, such as differing judgments by professionals in similar cases, which can lead to inefficiency and unfairness. Kahneman and his coauthors suggest that reducing noise is as important for improving decision-making as reducing bias. They propose strategies such as structured decision-making processes and the use of algorithms to make more consistent and accurate judgments.

The dual process theory provides a framework for understanding the mechanics of thought, highlighting the interplay between intuitive and rational processes. Kahneman's newer contributions shift the focus toward the quality and consistency of those decisions, emphasizing that variability (noise) in judgment is a significant issue in its own right. Although biases are about systematic deviations from rationality, noise points to the randomness in our decision-making processes. Kahneman's work consistently illuminates the complexities of human judgment, pushing for a broader understanding of what it means to make decisions and how those processes can be refined for better outcomes.[39]

Prospect Theory

Prospect theory, developed by Daniel Kahneman and Amos Tversky in 1979, is a foundational theory in behavioral economics that describes how people choose between probabilistic alternatives that involve risk, when the probabilities of outcomes are known. This theory is a critical departure from the expected utility theory (EUT), which had previously been the dominant paradigm for understanding economic decision-making under uncertainty. Prospect theory introduced several key concepts that have fundamentally changed our understanding of human behavior in economic contexts.[40]

At its core, prospect theory introduces several key concepts that explain why individuals often deviate from EUT. First, it posits that people evaluate outcomes as gains or losses relative to a reference point, typically their current situation, rather than considering outcomes in absolute terms. This idea contrasts with EUT, which assumes that people assess outcomes based on final assets. Second, the theory suggests that losses loom larger than gains, meaning the pain of losing is more intense than the pleasure of an equivalent gain. This asymmetry in emotional impact leads to risk-averse behavior when facing gains and risk-seeking behavior when facing losses, a phenomenon known as loss aversion. Another critical aspect is the concept of diminishing sensitivity, which suggests that the subjective value of changes in wealth diminishes as the amount increases, both for gains and for losses. This means that the difference in utility (or subjective value) between $100 and $200 is more significant than between $1 100 and $1 200, even though both scenarios involve the same absolute difference. Last, prospect theory accounts for probability weighting, when people tend to place excessive weight on small probabilities and lower weights on moderate to high probabilities, affecting their risk assessment. This leads to overestimation of unlikely events and underestimation of likely events.

Prospect theory has wide-ranging implications for economics, finance, psychology, and policymaking. It provides a more accurate framework for understanding decision-making in the face of risk, explaining phenomena that EUT cannot, such as why people buy both insurance (to avoid losses) and lottery tickets (overweighting small probabilities). These concepts collectively explain a wide range of human behaviors in economic decision-making, including why individuals might choose options that seem irrational from a purely economic standpoint.

The theory has been applied to various fields to explain behaviors like the equity premium puzzle in finance, when the returns on stocks significantly exceed the returns on safer bonds, contrary to what EUT would predict. It also sheds light on consumer behavior, such as the endowment effect (valuing owned items more than identical items not owned) and framing effects (the influence of presentation on decision-making).

In short, prospect theory has revolutionized our understanding of economic decision-making under uncertainty, providing a more nuanced and psychologically realistic account of human behavior than previous models. It highlights the importance of psychological factors in economic decisions, influencing a wide range of applications from policy design to financial market analysis.

Nudging

The journey to formulating nudging began with Thaler's observations of real-world behaviors that did not align with traditional economic models, which assumed that people are perfectly rational actors. Thaler's early research in behavioral economics demonstrated that people often make decisions based on cognitive biases and heuristics, leading to choices that deviate from what would be predicted by standard economic theory.

One pivotal moment came from his collaboration with psychologists, particularly Daniel Kahneman and Amos Tversky, whose work on cognitive biases and prospect theory significantly influenced Thaler. By integrating psychological insights into economic models, Thaler began to see how small and seemingly insignificant features of decision environments (which he later coined as *choice architecture*[41]) could have a profound impact on people's choices without coercing them. This understanding laid the groundwork for nudging.

Additionally, he recognized that the context in which decisions are made can significantly influence outcomes. Factors such as framing effects (how choices are presented), the physical environment, social norms, and cultural background can all shape decision-making processes. Thaler's insights into how choices can be influenced led him to acknowledge the potential for nudges – subtle changes in the way choices are presented to people that can guide their decisions toward more beneficial outcomes while preserving their freedom of choice. These nudges maneuver using known psychological mechanisms, such as loss aversion, status quo bias, and the power of defaults, to encourage better decision-making.

The concept was revolutionary because it offered a new way to think about designing policies, programs, and products. Nudging did not change the choices that improved people's lives in the long-term, but simply helped people make them. Thaler's work on nudging has since been applied across a wide range of areas, from public policy and health care to financial planning and environmental conservation, demonstrating its broad applicability and impact.

Perhaps one of the most influential applications of Thaler's work is in the domain of retirement savings. As previously mentioned, Thaler codeveloped the SMarT[42] program with Shlomo Benartzi, which nudges employees to commit in advance to allocating a portion of their future salary increases toward their retirement savings. This approach leverages several behavioral insights, including loss aversion and inertia, effectively increasing savings rates without reducing employees' take-home pay in the present.

Another significant example is the case of organ donation. The way options are presented can significantly affect organ donation rates. Thaler has highlighted the difference in organ donation rates between countries that use an opt-in system (in which individuals must actively choose to become donors) and those that use an opt-out system (in which individuals are presumed donors unless they choose not to be). The latter, which can be considered a nudge, has been shown to significantly increase the rate of organ donation, leveraging the power of defaults.

A classic example of how choice architecture and nudges can influence consumer behavior through pricing strategies involves the use of decoy pricing. This approach can significantly affect buying habits, particularly in how consumers perceive value and make choices between and among products or services.

Decoy pricing can be found, for instance, in subscription services, such as memberships for online platforms, magazines, or software. The company offers three subscription options: a basic version at $5 per month, a premium version at $15 per month, and a decoy option, which is a slightly less attractive premium version (e.g. fewer features than the premium but more than the basic) at $14 per month. The decoy is designed to make the full premium option appear more valuable in comparison, nudging consumers toward choosing it over the basic option.

The pricing structure is deliberately designed to highlight the attractiveness of the premium option by placing it next to a decoy that is only marginally cheaper but significantly less valuable. Consumers comparing the two premium options are nudged to perceive the full premium option as offering greater value for a minimal price increase. In other words, pricing structure is an element of choice architecture.

This pricing strategy exploits the consumer's tendency to compare options relative to one another. When presented with a decoy, consumers are more likely to choose the premium option over the basic one because the premium appears more beneficial in the context provided by the decoy. The decoy effectively nudges consumers toward a higher-priced option that they might not have considered otherwise, increasing the company's revenue per customer.

Decoy pricing works due to several psychological principles. The first is the contrast effect: The decoy makes the premium option look better in comparison, exploiting the contrast effect where the value is assessed relative to other available options. Second, the presence of a higher-priced option can serve as an anchor, making the price of the premium option seem more reasonable. And last, consumers are nudged to choose the option that seems to offer the best value for money, avoiding the potential regret of missing out on better features for a slightly higher price.

Decoy pricing is a powerful example of how choice architecture and nudges can be used to steer consumer decisions in the marketplace. By carefully designing the set of choices available, businesses can influence consumer perception of value and direct them toward preferred outcomes, demonstrating the subtle yet significant impact of pricing strategies on buying behavior.

In the context of retail environments, particularly grocery stores, the strategic placement of products acts as a powerful nudge, influencing consumer purchasing habits toward healthier choices through thoughtful choice architecture. By positioning fruits, vegetables, and other healthy foods at the front of the store or at eye level on shelves, retailers can leverage the visibility and accessibility of these options to affect shopping behavior. This approach takes advantage of the availability heuristic, where items that are immediately visible or encountered first are more likely to be chosen by consumers. Additionally, by presenting healthy options prominently, stores implicitly set a health-conscious tone for the shopping experience, nudging consumers toward making healthier purchases even if they had not specifically planned to buy these items beforehand.

Further enhancing the impact of product placement, grocery stores can employ signage and promotional materials that emphasize the health benefits of certain foods or the sustainability of local produce. This informational nudge works alongside the physical arrangement of products to cater to consumers' growing interest in health and environmental sustainability. Through these subtle yet effective modifications in choice architecture, retailers can significantly influence consumer behavior, demonstrating how nudges can encourage more beneficial choices without restricting consumer freedom.

Experimentation

"It does not matter how beautiful your theory is, it does not matter how smart you are. If it does not agree with the experiment, it's wrong."

– Richard P. Feynman

The origins of experimentation in economics and its pivotal role in the development of behavioral economics can be traced back to several key moments and figures in the history of economic thought. Experimentation in economics, particularly laboratory experiments, began to gain prominence in the mid-twentieth century, although the roots of experimental methods can be found earlier in the works of economists such as Vilfredo Pareto[43] and Irving Fisher,[44] who hinted at the value of experimental approaches in understanding economic behaviors.

The systematic use of experiments to study economic principles started in earnest with economists like Edward Chamberlin and Vernon Smith in the 1950s and 1960s. Chamberlin, for instance, conducted market experiments that challenged the traditional theory of perfect competition, showing how market imperfections could arise from individual behaviors.[45] I met Vernon Smith in my second year as a PhD candidate. He was regarded as the father of experimental economics and conducted controlled laboratory experiments[46] that laid the groundwork for understanding market operations, public choice, and property rights through an experimental lens. His pioneering work demonstrated the power of experiments in revealing the mechanisms of economic behavior, earning him the Nobel Prize in Economic Sciences in 2002.

The integration of experimental methods with psychological insights enabled behavioral economists to design experiments that not only tested economic theories under controlled conditions but also incorporated psychological realism. The 1980s and 1990s saw a surge in behavioral experiments that explored various aspects of economic behavior, from market transactions and negotiations to game theory and altruism.

Today, experimentation is a core part of behavioral economics, enabling researchers to test hypotheses about human behavior in economic contexts with precision and control. The use of field experiments, in addition to laboratory settings, has expanded the reach of behavioral economics into real-world applications, influencing public policy, finance, health care, and marketing. Experimentation has enabled behavioral economists to rigorously test interventions, such as nudges, that can lead to better outcomes for individuals and societies.

How It Works

An experiment in behavioral economics is a controlled study designed to test hypotheses about how people make economic decisions, particularly under varying conditions that simulate real-life scenarios. Unlike traditional economic experiments that often assume rational behavior among participants, behavioral economics experiments explicitly account for psychological factors, biases, and cognitive limitations that influence decision-making. The goal is to better understand the complexity of human behavior in economic contexts, going beyond the simplistic assumptions of rational choice theory.

Behavioral economics experiments can be conducted in both laboratory settings and natural environments, known as lab and field experiments, respectively. The process begins with the design phase, where researchers formulate hypotheses based on observations of human behavior. These hypotheses aim to explore various facets of decision-making, such as the effects of loss aversion, social norms, and how information framing influences choices.

In lab experiments, a highly controlled environment allows for the isolation and examination of specific behaviors through tasks or games, whereas field experiments take place in real-world settings, observing behaviors in their natural context but still under manipulated conditions to test hypotheses. The selection of participants is crucial, aiming to either reflect a broad population or target specific demographic groups to ensure the applicability or relevance of the findings.

Participants are randomly assigned to different conditions – varying in information, incentives, or available choices – to isolate the effects of experimental variables. This random assignment is key to attributing differences in outcomes directly to the manipulation of variables rather than external factors. Data collection encompasses not just the choices made but also might include reaction times, physiological responses, or self-reported attitudes, providing a multifaceted view of behavior under study. The analysis of this data, through statistical methods, determines the alignment of observed behaviors with initial hypotheses, offering insights into the underlying mechanisms of economic decision-making.

Behavioral economics experiments have revealed a wealth of information about how factors like emotions, social influences, and cognitive biases affect economic decisions. They have challenged traditional economic models by demonstrating that people often behave in predictably irrational ways, influenced by a complex array of psychological factors. This experimental approach has broadened the scope of economics, making it a more interdisciplinary field that incorporates insights from psychology, sociology, and neuroscience.

Uber and Experimentation

Uber, the global ride-sharing platform, has famously used experimentation and behavioral economics to optimize its operations and enhance driver engagement. One well-documented example involves Uber's use of targeted experiments to encourage drivers to stay on the road longer, especially during times of high demand. By analyzing vast amounts of data on driver behavior and ride patterns, Uber identified strategies to nudge drivers toward desired behaviors without overtly forcing their hand.

In one experiment, Uber tapped into the power of goal setting and instant feedback. The company found that drivers often set earnings goals for themselves each day. Leveraging this insight, Uber began sending messages to drivers who were close to hitting their earnings target, encouraging them not to log off yet. This simple nudge was based on the premise that the prospect of reaching a goal would motivate drivers to work a little longer. This experiment also drew on loss aversion, suggesting that the pain of falling just short of a goal is a strong motivator.

Another example of Uber's experimentation involved dynamic surge pricing. By adjusting fares in real-time based on supply and demand, Uber could nudge drivers toward areas with higher demand or times of day when fewer drivers were on the road. Drivers seeing higher potential earnings in certain areas or times were more likely to adjust their schedules accordingly, helping balance supply and demand across the network. This approach not only increased drivers' potential earnings but also improved service reliability and wait times for riders.

It is important to note that Uber's extensive and effective use of experimentation has generated its fair share of criticism, with the *Harvard Business Review* even dubbing their use "how *not* to apply behavioral economics."[47] Critics point out that Uber can nudge drivers to work at hours and locations that, although more lucrative for the company, might be less beneficial for the drivers. This discussion is part of a larger, ongoing debate on the nature of nudging – whether "good" or "bad" nudges exist, and whether these subliminal and noninvasive strategies are a form of manipulation.[48] Such controversies boil down to fundamental conceptions of free will and persuasion, cementing behavioral economics not only as an incredibly powerful tool but also one requiring a delicate balance of consideration.

Dan Ariely once collaborated with a major consumer goods company to explore the effects of "free" offers on consumer behavior, using principles of behavioral economics. The partnership aimed to create multiple websites with varying offers and pricing to observe customer reactions and gather data beneficial for both academic research and the company's marketing strategies. However, ethical concerns arose when a marketing team member highlighted potential inequalities among customers due to the experiment, comparing the trade-offs to those in clinical trials but with less at stake. Despite Ariely's argument that experimentation is crucial for obtaining evidence beyond mere intuition, the project was ultimately canceled due to the company's reluctance to potentially disadvantage any customers.[49]

This experience illustrates a broader reluctance within the corporate world to engage in experimentation, despite its potential to provide valuable insights. Ariely observed that companies often prefer to rely on consultants' advice or focus groups rather than conduct experiments that might reveal more accurate information about effective strategies. This aversion stems from a dislike of short-term sacrifices for long-term benefits and a preference for the false security of expert opinions over the uncertainties of experimental outcomes. Nevertheless, Ariely notes that some companies, like Intuit under Scott Cook, are embracing a culture of experimentation, recognizing its value in generating evidence that can guide better decisions than intuition alone. This shift toward acknowledging the importance of evidence over intuition marks a slow but significant change in how businesses approach decision-making and strategy development.

CHAPTER

3

Minimal Data, Maximal Impact
From Big Data to Minimum Viable Data

"It is a capital mistake to theorize before one has data. Insensibly, one begins to twist the facts to suit theories, instead of theories to suit facts."

– Sherlock Holmes

"Sometimes, I miss the simplicity of particle physics," I mentioned to a friend, noticing his puzzled look. "At CERN, everything was more black and white. We had massive amounts of data and specific theoretical models to test. The questions were direct: What happens if the Higgs boson isn't found? Or what if we find more than one? Though these scenarios were complex, our approach was clear: just keep accelerating protons at higher energies."

Seeing that my point wasn't quite hitting home, I added, "Compare that to business environments where the challenges are often vague. The necessary data might be scattered across different systems, and that's if you are lucky. More often, the right data is not collected at all, is biased, or just insufficient for answering critical questions. By contrast, at the ATLAS experiment,[1] we had a definite goal: accelerate 10^{25} protons to nearly the speed of light, colliding them every nanosecond.[2] It was a daunting task, but we were precisely equipped to gather the data needed to explore phenomena mere moments after the big bang." I watched him nod, starting to connect the dots. Even the most difficult challenges are solvable with targeted solutions, but you simply cannot find the answers to the questions you do not know.

I found my way into the field of data and AI serendipitously, fortunate to straddle the intersection of physics and economics. This unique confluence has enriched my perspective, enabling me to pose unconventional questions and devise innovative solutions. Before I delve into how I leverage insights from these disciplines to affect your organization, let us understand why this fusion is crucial in today's landscape.

Our relationship with the world has changed an immense, almost unfathomable amount throughout the twenty-first century. Emerging technologies have revolutionized our connections and the products we cherish. Every day, enterprises generate millions of terabytes (TB) of data, yet despite this abundance, many struggle to pinpoint value. The greatest opportunities for growth and competition are too often lost in a sea of data. This gap in accuracy between the questions asked and the data needed to answer them leaves customers disconnected from the products they love and use, blocking companies from realizing meaningful growth.

The challenge is multifaceted: an overemphasis on technology for its own sake, a maze of available data with no clear path, and a misalignment between technical potential and strategic business goals. These factors complicate the ability to tap into the full potential of data and truly drive enterprise growth.

In my discussions with senior executives at Fortune 500 companies, I often encounter a litany of concerns about data use: "We have so much data, but we don't know where to start to monetize it" or "We receive conflicting advice from different consultants on our data strategy." Questions also arise about the adequacy of existing data, the potential of synthetic data, and the trustworthiness of AI-generated information: "Do we have enough data? Can we enhance our data with external datasets? Should we consider synthetic data, and can we rely on it if it's AI-generated? What about potential biases?" These inquiries are just the tip of the iceberg.

In this chapter, my aim is to clarify these complex issues surrounding big data. I believe that by exploring key concepts such as minimum viable data (MVD), synthetic data, and biases, and by providing practical examples of how machine learning (ML) and AI models are applied, this discussion will significantly conserve time and resources for any organization, particularly in the realms of data, AI, and digital transformation strategies.

By highlighting the connections between essential business drivers and the factors influencing customer decisions, large enterprises can cut through the clutter. Asking the right questions and harnessing data effectively will unlock innovation and drive strategic advancements.

How Much Data Are We Talking About? Lots and Lots

One of the advantages of my early career at CERN, particularly on the ATLAS detector when I was 22, was the sheer volume of data we dealt with daily – it was simply a fact of life. I vividly remember the first significant dataset we managed, which was 7 TB. Over lunch, we'd casually discuss the number of "cores" needed to run our "jobs." However, the real insight lies not in the volume of data but in how we used it. We did not attempt to clean all 7 TB or even the 20 TB we received the following day – our focus was sharply defined. We knew exactly which features and signals were crucial for the models we aimed to build and test against our hypotheses.

The processing logic was intricately coded into a digital, pipelined system with multiple stages, primarily using field-programmable gate arrays (FPGAs). These devices executed programmable algorithms in parallel, maintaining a fixed latency, capable of handling approximately 300 gigabytes per second of input data.[3] This setup wasn't just about managing data; it was about targeting the right data effectively to confirm or refute scientific theories.[4]

The key takeaway from my experience at CERN was that it wasn't necessary to process all the data; we just needed the right data. We had effectively reverse-engineered the data pipelines based on the models we aimed to construct and the hypotheses we planned to test. There would be no Nobel prizes for merely organizing and cleaning TB of data! Instead, our approach was targeted: we extracted relevant signals that were directly guided by our specific scientific inquiries.

This targeted flow is precisely what I observe lacking in many enterprises today. Typically, organizations begin with the "universe of data" – focusing on what data is collected, needs cleaning, organizing, and governance. However, a more effective approach starts with the "universe of questions." This involves evaluating hypotheses, identifying how different use cases connect, and reverse-engineering to determine which data is essential for validating these hypotheses and answering critical questions.

In essence, although many start with data, the challenge is not just a data problem. The fundamental issue stems from failing to ask the right questions or to prioritize use cases that enable the organization to build a robust foundation. This foundation is crucial for addressing multiple queries, reducing duplication and redundancy, and enhancing efficiency and the speed of adoption. Clear questions and well-defined hypotheses are potent tools that guide targeted data collection and streamline the entire process.

As of April 2023, our world was producing approximately 2.5 quintillion bytes of data daily,[5] demonstrating the vast scale of the digital era's growth. This data comes from myriad sources – social media, digital media, transactions, business operations, and Internet of Things devices – fueled by rapid digitization, expanding internet access, and the proliferation of smart technologies. Daily, we generate an astonishing 328.77 million TB, summing up to about 120 zettabytes (ZB) annually. With projections pointing to an increase to 181 ZB by 2025, this trend exemplifies not just growth but an exponential surge in data creation.[6]

This staggering increase – in which 90% of the world's data has been generated in the past two years alone – showcases the relentless pace of the digital age and its potential for future expansion. The anticipated 150% rise in data by 2025 further highlights the relentless evolution of technology and its increasingly significant impact on our lives. As we delve deeper into this data-driven era, the real challenge – and opportunity – lies in effectively leveraging this vast volume of information. It begins with crafting precise hypotheses, posing the right questions, and targeting specific data through refined management and analytical techniques. This approach will not only optimize the use of this enormous data reservoir but also propel innovation, enhance decision-making, and advance our integration with technology.

The path to harnessing the power of big data is marked not by the quantity of data gathered but by the quality of questions asked and the precision with which data is used. By focusing on targeted data collection and analysis, organizations can unlock significant value, driving strategic growth and operational efficiency in an increasingly data-saturated world.

You Do Not Need a Lot of Data to Get Started, You Need the MVD

The concept of MVD is inspired by the lean startup[7] methodology's minimum viable product (MVP). In the realms of data science and AI, MVD denotes the smallest dataset required to effectively train a ML model to a specified performance standard. This idea of "smart sizing" data has been pivotal in the shift toward data-centric AI, where the focus is on the quality, relevance, and cleanliness of data rather than its quantity. This approach advocates for a more efficient and focused method of data collection and preparation.

Andrew Ng,[8] a pioneer in the development of deep learning and artificial intelligence (AI), champions this "data-centric AI" approach, prioritizing data quality over complex AI algorithms, which are now more accessible than ever. Ng argues that the key to effectively leveraging AI lies in the meticulous selection, preparation, and governance of data. By employing strategies that optimize data use – requiring less but more precise data – businesses can develop efficient AI systems that compete with those of tech giants, even with smaller datasets. Ng's emphasis on data consistency and strategic curation of training sets signals a significant shift toward more intelligent, accessible AI development. In this model, continuous data refinement and system retraining are crucial for success in AI.

Embracing MVD enables organizations to use resources more efficiently, shorten data processing times, and accelerate the iterative cycles of model training and refinement. MVD is not just about reducing data size; it's about intelligently identifying which data elements are most crucial for model performance and decision-making. This streamlined approach facilitates quicker deployment and agility in adjusting models based on initial insights and performance feedback. For example, in a customer sentiment analysis tool, the MVD might be recent reviews and ratings, focusing on keywords and sentiments rather than extensive historical transactional data or detailed customer profiles. In predictive maintenance for manufacturing, MVD would likely include recent operational data and error logs, emphasizing timeliness and specificity over comprehensive historical records.

MVD promotes a focus on data quality – accuracy, consistency, and relevance to the problem at hand. Starting with high-quality, targeted data can significantly enhance model performance, even with a smaller dataset. This initial model serves as a foundation, similar to an MVP, which can be expanded and refined with additional data based on user feedback and emerging needs.

By identifying the MVD, organizations make strategic decisions about data acquisition and curation, guiding them on what types of data are most valuable and what additional data might be necessary to tackle specific challenges or enhance model accuracy. Through these practices, MVD emerges as an approach that balances efficiency with effectiveness, enabling smarter data practices focused on achieving substantial, meaningful results.

Asking the Right Questions, Again!

Asking the right questions is critical in defining an optimal MVD set, ensuring that the focus is tightly aligned with the intended outcomes of a project. This process begins with a thorough understanding of the problem statement and the specific goals of the model. Such clarity helps in discerning which data elements are essential and which are superfluous. Teams should consider the key predictors of the outcome, the availability and quality of the data, the necessary granularity and timeliness of the data, and the cost implications of data acquisition and maintenance. Prioritizing data that provides the greatest predictive power and relevance to the task at hand guides the selection process, emphasizing the most impactful variables and eliminating unnecessary information. This method streamlines data collection and processing efforts.

This targeted approach to defining MVD not only speeds up the development and deployment of ML models but also ensures these processes are constructed on a data foundation most indicative of the desired outcomes. Regularly revisiting these questions facilitates the iterative refinement of the dataset, enabling it to adapt to changes in project goals, emergent insights, or shifts in the external environment. As a result, the MVD evolves over time, staying aligned with the project's objectives and the dynamic nature of data. By strategically focusing on the right data from the beginning, organizations can improve the efficiency, effectiveness, and economic viability of their data-driven initiatives, leading to more precise decision-making and enhanced outcomes.

A few years ago, about six months after launching Theory+Practice, I had the opportunity to collaborate with one of the largest logistics companies in the world. The marketing team, focusing on e-commerce and retail marketing and customer engagement, was tasked with identifying AI and ML use cases and gathering the necessary data. During a meeting in Memphis, I learned they had pinpointed 120 different use cases, ranging from customer segmentation to advanced recommendation systems and algorithms for suggesting the next best action. The team had also identified 25 different datasets.

Sitting there, I was puzzled by the lack of clear prioritization criteria, value metrics, or defined return-on-investment (ROI) goals that should have guided the ranking of these use cases. It struck me as inefficient: Without knowing which use cases to prioritize, how could we possibly determine the most appropriate datasets? Additionally, there were concerns about identity resolution and the ability to create a comprehensive 360° view of customers across different datasets – one might capture online interactions, another could reveal insights into promotional and price sensitivities, and a third might inform us about customer preferences regarding communication methods and timing.

Despite the team's diligent effort in identifying these use cases, there was no clear road map or guiding principles to ensure value creation, increase efficiency, and reduce duplication. The necessary connections were not being made to ensure that investments in the data foundation would maximize the addressable use cases, rather than perpetually starting from scratch with new initiatives.

This lack of strategic foresight in the initial stages of project planning presented a significant barrier to leveraging the full potential of AI and ML technologies. The absence of a structured road map not only hindered the team's ability to effectively use the identified datasets but also complicated the integration of new technologies into existing workflows. The realization of these challenges led to a fundamental shift in our approach.

We initiated a comprehensive review of the use cases and datasets to establish a hierarchy based on potential impact and feasibility. This involved setting clear, measurable objectives for each use case, aligning them with overarching business goals, and identifying key performance indicators to track progress and outcomes. We also emphasized the importance of data integration, ensuring that each dataset could be harmoniously linked to provide a unified and complete view of customer interactions.

Through these adjustments, we aimed to create a more efficient and targeted strategy that not only reduced redundancy and waste but also enhanced the overall effectiveness of the team's efforts. By prioritizing use cases with the highest potential for ROI and ensuring a cohesive data strategy, the organization could better align its technological investments with its strategic objectives, paving the way for more informed decision-making and robust customer engagement strategies.

The right questions, rooted in a deep understanding of the underlying problems and challenges, illuminate potential avenues for both value creation and capture. These inquiries facilitate the identification of connections among various use cases, fostering the development of a strategic road map. This road map is designed to enhance efficiency and minimize redundancies by pinpointing the most relevant datasets that maximize the predictive and analytical capabilities of statistical and ML models. It's crucial to recognize that our ultimate goal is to influence specific behaviors or to make more informed decisions. This goal is closely linked to the insights gleaned from data and the quality of data employed in the models and algorithms. A rigorous approach to data selection and use directly strengthens the impact and efficacy of our data-driven initiatives, ensuring that they are not only aligned with but also advance the organization's strategic objectives.

MVD also has a wide range of applications in retail, because development of demand sensing and forecasting models requires the identification and prioritization of crucial data elements. These elements must provide the necessary insights to enable accurate predictions while maintaining minimal initial complexity. The objective is to begin with a simplified model that can be enhanced iteratively over time. Demand-sensing models, aimed at responding to short-term market changes, integrate immediate data sources such as recent sales figures, current inventory levels, ongoing promotional activities, and relevant external factors like weather conditions or local events. This enables retailers to quickly adjust to market fluctuations, providing accurate short-term demand predictions.

Conversely, demand forecasting models, which support long-term planning, necessitate a broader but selectively refined dataset. These models typically rely on historical sales data, ranging from several months to years, complemented by product details, general market trends, and macroeconomic factors that influence consumer behavior over time. Initiating with these fundamental data points enables businesses to establish a baseline model that is both operational and adaptable. As the model matures, more nuanced data such as customer demographics and supply chain details can be integrated, steadily improving the model's precision and its relevance for forecasting future demand.

In both scenarios, the concept of MVD serves as the initial foundation. Once the preliminary models are constructed and validated, further data can be methodically added to refine and enhance the predictions. The essence of MVD lies in starting with the most significant data to swiftly create functional models, then methodically enlarging the dataset based on evolving needs and insights derived from earlier analyses. This approach ensures that retail businesses remain agile and responsive, leveraging a data-driven strategy to effectively anticipate and meet consumer demands.

Synthetic Data: What It Is and What It Isn't

In initial discussions with executives from large enterprises, the topic of synthetic data frequently emerges. Many leaders express concerns about whether their internal data suffices for producing high-quality models, citing issues related to the volume and quality of their data. The daunting task of organizing, synthesizing, and cleaning relevant data often leads to inquiries about alternative datasets. Although I emphasize that external datasets are only beneficial in specific scenarios and that their own data captures critical nuances of their business, the conversation typically shifts to synthetic data.

What exactly is synthetic data? It is generated by generative AI models trained on real-world data samples. These algorithms learn the patterns, correlations, and statistical properties of the data. Once trained, the generators produce synthetic data that is statistically indistinguishable from the original but devoid of any personal information. This makes synthetic data ideal for training ML models and providing privacy-safe versions of datasets for data sharing.

Synthetic data, not derived from real-world events, is crafted via algorithms or simulations to mimic the statistical properties of authentic data. This type of data is invaluable in situations where actual data might be limited, sensitive, or difficult to obtain. The creation of synthetic data allows for robust data analysis, model training, and system testing without compromising privacy or breaching data protection laws.

Synthetic data is a pivotal element in contemporary data analysis, ML model training, and testing, providing a robust solution to issues of data privacy, sensitivity, and scarcity. It comes in various forms, each crafted to meet distinct needs and scenarios. Fully synthetic data is generated exclusively through algorithms, creating completely new datasets that mirror the statistical properties of real data. This makes it especially suitable for testing and development in contexts where using real data could raise privacy or ethical concerns. For example, financial institutions employ this type of data to develop and refine fraud detection algorithms[9] without risking customer privacy.

Partially synthetic data modifies only the sensitive components of real datasets. This approach is frequently used in health care research[10] to anonymize patient details while maintaining the medical data's integrity for thorough analysis and study.

Further expanding synthetic data's utility, hybrid approaches mix real and synthetic elements to enhance datasets or simulate conditions absent in the original data. This enables organizations to project outcomes in untested markets or new product introductions. Additionally, agent-based synthetic data, produced through simulations of agent interactions, offers insights into complex systems such as urban traffic patterns. This aids significantly in informed decision-making within urban planning and public policy.[11]

Each type of synthetic data serves distinct needs, from enhancing privacy and security to augmenting existing datasets for better model training. The choice of synthetic data type depends on the specific requirements of the project, such as the need to protect sensitive information, the requirement for large volumes of training data, or the exploration of hypothetical scenarios not represented in the real data. As such, synthetic data has become a valuable asset in data science, AI development, and beyond, enabling innovation while navigating the constraints of data availability and ethical use.

These varied types of synthetic data collectively enable organizations across multiple sectors to leverage the power of data analytics and AI, driving innovation while carefully addressing privacy and data availability issues. Through its strategic application, synthetic data not only improves model accuracy and testing but also facilitates new research and development opportunities, free from the ethical and practical constraints associated with using real-world data.

If you want to get very geeky, let us dive into the intricate world of data types! It's essential to clarify the distinction between AI-generated synthetic data and traditional mock data – a common source of confusion. Unlike mock data, which is typically created randomly or follows predefined rules without relying on real data samples, AI-generated synthetic data originates from actual datasets. This is achieved through generative AI models that analyze real data to produce new datasets that maintain the original's statistical characteristics, enabling more accurate and realistic simulations. This differentiation also applies to structured versus unstructured synthetic data. Structured synthetic data, for example, appears in formats such as financial records or customer relationship management databases and often includes behavioral or time-series data.

Furthermore, it's important to distinguish synthetic data from Monte Carlo data, both of which are used in simulations and modeling across various fields such as finance, health care, and ML but serve different purposes and are generated through distinct processes. Synthetic data is artificially created to mimic the statistical properties of real-world data and serves as a substitute in situations when using actual data is impractical, costly, or sensitive – such as in software testing, training ML models without revealing personal information, or research in which real data might not be available. It is generated through algorithms or models that learn from and replicate a dataset's characteristics, ensuring that the synthetic version preserves the structure, relationships, and variability of the original data without exposing individual data points.

Monte Carlo data, however, is produced using Monte Carlo simulations – a method that models the probability of different outcomes in processes that cannot be easily predicted due to random variables. This technique is extensively used in fields like finance for risk assessment and portfolio management, physics for studying particle interactions, and operations research for optimizing complex systems. Monte Carlo simulations involve repeated random sampling to achieve numerical results, running a model thousands or millions of times with varying inputs to simulate a broad range of outcomes. The resulting data offers statistical distributions of potential outcomes, aiding in understanding risks, uncertainties, and the probability of various scenarios.

Although both synthetic and Monte Carlo data are artificial and used in simulations, synthetic data aims to replicate the properties of real-world data without using actual data points, primarily for privacy, testing, or when real data is not accessible. Monte Carlo data is specifically generated through repeated random sampling to model probabilities of various outcomes in processes influenced by uncertainty, widely used for risk analysis and decision-making under uncertain conditions.

Survey Data to the Rescue

In 2023, we worked with one of the largest beverage companies in the world on a very interesting problem: The marketing department was interested in understanding drivers of behavior among different cohorts of consumers. They wanted to know what makes millennial versus Gen Z tick, or what the differences between the northwest of the United States and the southeast are, what makes them develop a stronger sense of affinity to a brand or a product. What made it difficult, like many other consumer packaged goods companies, they do not have firsthand data from consumers. They have access to huge amounts of shipment, product, and some transactional data from their partners and grocer, but no information on customer demographics or preferences.

To tackle this challenge, we decided to tap into survey data that can serve as a rich source of insights into consumer habits and behaviors. Using ML models and statistical analysis, this data can be analyzed through the lens of behavioral economics to understand the underlying drivers of economic decisions. This relationship is bidirectional; not only can behavioral economics theories help interpret survey data but the data itself can also provide empirical evidence to test and refine those theories.

Survey data usually consists of information collected from respondents through questionnaires, aiming to gather insights on a wide array of subjects such as consumer behavior, opinions, and demographic traits. This data is crucial for understanding target demographics, identifying market trends, and gauging societal attitudes. To ensure the high quality and minimal bias of survey data, it's essential to use clear, neutral language in question formulation, secure a diverse and representative sample of the target population, maintain respondents' anonymity, pre-test the survey for potential issues, and keep the survey concise to prevent respondent fatigue.

Our survey data spanned a few years, and there were thousands of individuals who had answered the same set of questions about the product and the brand over years. We grouped behaviors and the desired outcomes into intuitive categories; for example, we wanted to understand how an individual's sense of worth affects their affinity with the brand, or whether the sense of energy and happiness or connecting with friends increased or decreased their consumption. Was price a big factor in their decision? Could we find strong enough correlations and predictive signals in this data? In other words, we wanted to use survey responses to predict outcomes or behaviors. For instance, responses to our survey on consumer habits could be used to train a model on predicting future purchasing behaviors based on demographic data and history. At the end, we managed to build over 500 ML and statistical models shedding light on strong correlations and predictive factors behind behaviors of interest. These models uncovered $1.2 billion worth of opportunity by identifying and prioritizing the most significant drivers of behavior. All this was possible by carefully integrating survey data into ML models through proper feature engineering – controlling for factors that would remove the usually inherent bias in survey data.

It is important to note that incorporating survey data into ML models requires careful preprocessing, such as encoding categorical data, handling missing values, and normalizing responses. Continual evaluation of the model's performance and bias is also vital, adjusting the training process as necessary to ensure that the complexities and nuances of the survey data are accurately reflected.

Another example of using survey data is in studying and uncovering how a new technology will spread and diffuse in the society. There is a large literature in economics devoted to this subject and many are familiar with the *S*-shape curve of diffusion with early adopters, early majority, and late majority, and so on. Many companies and organizations want to understand how their new products would do and what kind of marketing and to whom is going to make this product launch more successful. However, governments want to understand how citizens would react to a particular policy change or how fast they will adopt a new service.

What makes it difficult often is lack of relevant or appropriate data: obviously when a new product is introduced in the market, there are no historical purchase patterns for that product and usually data from similar or substitute products is used to forecast consumer demand or interest.

In 2020, almost a year into the COVID-19 pandemic and a few days before Christmas, I got a call from a government official explaining a dilemma: The government authorities in this province were in the process of introducing digital identity (DI). They had mapped more than 50 use cases that span across educational, financial, health, and government-related application of digital identity. However, they were faced with a question that they did not know how to approach: How should they prioritize these use cases and where should they start to roll out digital identity to maximize the speed of adoption by the general public? Should they use vaccination cards or are driver licenses a better place to start? Some countries started with bank cards and integrated DI into the financial system; should they do the same? Is it possible to predict the speed of diffusion at all? How to capture network and spillover effects?

These are indeed important and difficult questions, usually requiring two to three years of time-series data to estimate components of the *S*-shaped diffusion curve or the Bass model. Well, we did not have that kind of data because the program has not yet rolled out. But it was very important to at least find directional answers to the questions.

My team and I had less than a week to identify an approach and explore its feasibility. We got to work very quickly: one thing was obvious, we either had to identify an older "new technology" that had many similarities to DI that had been released in the same or similar region and had used historical data to estimate the speed of diffusion or find some new data. That's when we had an idea: How about building a survey that can extract citizen's preferences and their perceptions of how quickly this technology would diffuse in the population? Using survey data, not only could we get a better sense of the benefits and risks that citizens perceive but also we could extract time preferences and estimate the dynamics and speed of diffusion.

That's exactly what we did: using a particular survey design, we collected over 15,000 responses in a month and although the survey was less than 30 questions, we had collected very rich data on demographics, preferences, risks, and benefits as it related to many of the use cases this government cared about. We incorporated this data into ML models and statistical models that estimated the speed of diffusion. For example, we were able to predict that starting with a financial use case in combination with a particular health use case would increase the speed of diffusion four times over starting with a simple driver's license use case and would leave to more than 80% of society to adopt DI in four years. This would unlock a $25 billion opportunity for the government with about a $2 billion investment. We also saw how trust in government or tech giants also had a significant impact in the adoption of DI.

Using survey data and qualitative methods alongside traditional quantitative analysis can significantly enhance the predictive capabilities of ML models. This integrated approach allows for a deeper understanding of the underlying phenomena behind the data, incorporating human insights and complexities that quantitative data alone might miss, and involves a careful balance of art and science. By combining survey analysis with behavioral economics, researchers and practitioners can develop more nuanced strategies for influencing consumer behavior, designing products, and crafting policies that align with actual human behaviors and preferences.

CHAPTER

4

Building Intelligence
AI and ML Essentials, Transforming Data into Intelligence

> "Artificial intelligence, deep learning, machine learning – whatever you're doing, if you don't understand it – learn it. Because otherwise, you're going to be a dinosaur within three years."
>
> – Mark Cuban

One of my favorite moments in meetings with executives is when a senior executive, usually a C-suite leader, asks me to clarify the difference between artificial intelligence (AI) and machine learning (ML), or whether what we are doing is truly AI or just ML. I relish this question because it signals genuine curiosity and a willingness to understand the nuances rather than merely parroting jargon from previous consultants.

Their engagement indicates a desire to move beyond buzzwords to grasp the underlying principles.

In these moments, I take this opportunity to demystify the distinctions with clear examples, explaining classical AI, ML algorithms, generative models, and more. My goal is not to turn them into AI experts but to provide enough clarity so they can ask the right questions and foster a culture of inquiry among their teams. This understanding is crucial, especially given the explosive interest in generative AI, where making informed decisions can protect them from costly investments in the wrong technologies or solutions.

AI – and particularly its latest frontier, generative AI – has become a focal point in corporate boardrooms, leadership discussions, and even casual workplace conversations among employees eager to boost their productivity. However, beneath the aspirational headlines and the tantalizing potential lies a sobering reality: most AI projects fail. By helping executives understand the intricacies and ask the right questions, we can navigate this challenging landscape more effectively, steering toward successful AI implementations that truly add value.[1]

I saw a funny cartoon recently on LinkedIn titled "The Fastest Things on Earth." There was a picture of a cheetah, followed by an airplane, speed of light, and finally "people becoming experts in AI"! This is one of the reasons why we hear statistics such as 88% of AI projects fail to deliver positive return on investments[2] or 80% of AI projects never get out of the proof-of-concept phase – almost double the failure rate of corporate information technology projects from the mid-2010s.[3] But the issue of investing in the wrong AI project goes beyond these alarming numbers. Gartner predicted that through 2022, 85% of AI projects would deliver erroneous outcomes due to biases in data, algorithms, or the teams managing them. Other contributing factors include lack of

Building Intelligence

adoption and trust, insufficient experimentation and proper evaluation of data and models, and inadequate management and maintenance of AI projects.[4]

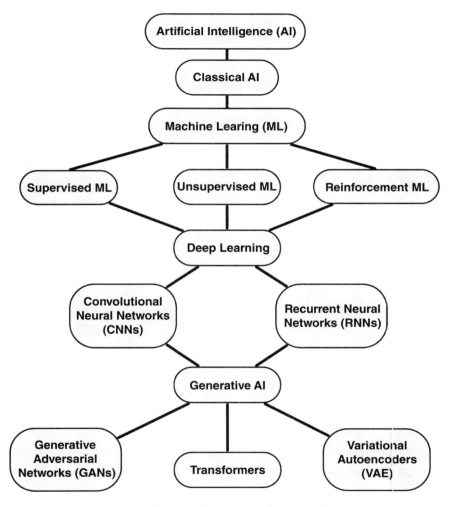

FIGURE 4.1 Overview of AI and ML: This diagram illustrates the hierarchical structure of AI, from classical AI to ML, deep learning, and generative AI, highlighting key subfields such as supervised and unsupervised learning, reinforcement learning, CNNs, RNNs, GANs, transformers, and VAEs.

In this chapter, my goal is to delve into the key differences among various ML algorithms and to highlight the scenarios and situations in which each can be most effectively applied. By understanding these distinctions, we can better navigate the complexities of AI, ensuring that our investments yield meaningful, reliable outcomes and drive real value for our organizations (see Figure 4.1 for an overview of AI and ML subfields).

Classical AI

Classical AI, also known as symbolic AI or good old-fashioned AI, refers to an approach to AI that was dominant from the 1950s to the 1980s that was characterized by the use of symbolic representations of problems and logical reasoning to solve them. These models are rule-based and rely on explicit programming of knowledge and rules by human experts.

To truly grasp the essence of classical AI, imagine stepping into a world where human intelligence is systematically translated into machine operations through well-defined rules and logical structures. Classical AI embodies the pioneering principles of AI, relying on symbolic representations and rule-based systems to emulate human cognitive functions. They use formal languages to encode knowledge about the world in a structured way, often in the form of if-then rules and apply logical rules to the knowledge base to infer new information or make decisions. In contrast to modern AI, which leverages vast datasets and neural networks, classical AI is built on a foundation of logical reasoning, search algorithms, and emphasizes the use of explicit knowledge encoded in a form that machines can process. Additionally, given the same input, they will always produce the same output, because the rules are explicitly defined and the reasoning process can be easily traced and understood and it follows a clear set of rules.

Building Intelligence

One of the quintessential applications of classical AI is the development of expert systems. These systems are designed to mimic the decision-making abilities of human experts in specific domains. Take, for instance, MYCIN, an early expert system developed in the 1970s to diagnose bacterial infections and recommend antibiotics. MYCIN was one of the first expert systems that used backwards chaining to identify potential diagnoses and prescribe medications.

MYCIN used a vast knowledge base of medical information and a set of inference rules to analyze patient data, much like a human doctor would. By applying logical rules to the input data, MYCIN could provide diagnostic recommendations with impressive accuracy for its time.[5]

Another fascinating example of classical AI is found in game-playing programs. Consider Deep Blue, the chess-playing computer developed by IBM. Although often associated with modern AI, Deep Blue's approach was heavily rooted in classical AI techniques. It relied on a combination of brute-force search algorithms and heuristic evaluation functions to explore possible moves and predict outcomes. By systematically evaluating a vast number of potential game states, Deep Blue famously defeated world chess champion Garry Kasparov in 1997, showcasing the strategic power of classical AI.[6]

Classical AI's prowess in logical reasoning is perhaps best exemplified by automated theorem proving. These systems are designed to prove mathematical theorems by applying formal logic rules. One notable example is the resolution principle, developed by Alan Robinson in the 1960s, which became a cornerstone of automated reasoning. Programs like Prolog, a logic programming language, were created to solve complex problems by defining relationships and rules. Prolog could, for instance, solve puzzles, perform natural language processing tasks, and even help in legal reasoning by logically deducing conclusions from a set of premises.[7]

Despite its early successes, classical AI faced significant limitations. Its reliance on explicit knowledge representation made it brittle and difficult to scale. The complexity of real-world problems often required an impractical number of rules and immense computational resources. As a result, the field experienced a shift toward ML and neural networks, which could handle unstructured data and learn from examples rather than relying solely on predefined rules.

In today's landscape, understanding classical AI provides valuable insights into the evolution of AI. It teaches us the importance of structured knowledge and the power of logical reasoning, reminding us that even as we harness the potential of neural networks and big data, the foundational principles of AI remain relevant. By appreciating the journey from classical AI to modern ML, we can better navigate the complexities of developing intelligent systems that are both powerful and explainable. Classical AI laid the groundwork for the field of AI. Although it might have given way to more flexible and scalable approaches, the lessons from classical AI continue to shape the future of intelligent systems, guiding us toward creating AI that not only performs tasks but also understands and reasons about the world in a human-like manner.[8]

ML

ML stands at the core of modern data analytics and AI, empowering systems to learn from data and improve over time without being explicitly programmed. It moves beyond traditional statistical methods, enabling more dynamic and adaptive solutions to complex problems. ML is a subset of AI involving the development of algorithms that enable computers to learn from and make predictions/decisions based on data. The quality and quantity of data significantly affect the performance of the models. The goal of ML models is to generalize to new, unseen data, not just to perform well on the training data. Once trained in iterative processes, ML models can automate decision-making processes, often outperforming humans in speed and accuracy.

Feature engineering is the process of using domain knowledge to create new input features from the raw data that help improve the performance of ML models. Feature engineering is a crucial step in the ML process, and is the process of using domain knowledge to create new input features from the raw data. To do this, raw data is transformed into informative features that improve the performance of predictive models. This involves creating new features, selecting important ones, and encoding data in a way that ML algorithms can use effectively. This process occurs within the machine more than relying on preprogrammed rules, so a machine can perform tasks without deterministic, explicitly defined instructions.

Suppose you are developing a model to predict equipment failures in a factory. The raw data might include temperature readings, vibration levels, and operational hours. Through feature engineering, you could create features such as the rate of change in temperature, the average vibration over time, and the ratio of operational hours to maintenance hours. These engineered features can provide more insightful information, leading to more accurate predictions.

Alternatively, consider a model designed to assess credit risk for loan applicants. The raw data might include income, age, and loan amount. Feature engineering can enhance this by incorporating features like debt-to-income ratio, credit history length, and the number of recent credit inquiries. These features provide a more comprehensive view of an applicant's financial health, thus improving the model's predictive power.

ML is an iterative process. Models are continuously refined based on new data and feedback. An initial model might start with basic features, and as more insights are gained, additional features are engineered, and the model is retrained. This iterative cycle of learning and improvement is what drives the effectiveness of ML applications.

At its heart, ML involves the development of algorithms that can identify patterns in data and make decisions or predictions based on those patterns, and there are several types of ML. In the supervised learning approach, the model is trained on a labeled dataset, which means that each training example is paired with an output label. The algorithm learns to map inputs to outputs, making it suitable for tasks such as classification and regression. For instance, a supervised learning model can be trained to classify emails as spam or not spam based on historical data of labeled emails.

However, unsupervised learning models are given data without explicit instructions on what to do with it. The goal is to find hidden patterns or intrinsic structures within the data. Clustering, segmentation, and dimensionality reduction are common techniques in unsupervised learning. A practical example is customer segmentation, in which an algorithm groups customers based on purchasing behavior without prior labels.

Reinforcement learning (RL) is another example of an ML system that involves an agent that learns by interacting with its environment and receiving feedback in the form of rewards or penalties. It's particularly powerful in scenarios where decision-making is sequential, such as game playing or robotic control. For example, RL can be used to train a robot to navigate through a maze by rewarding it for each step that brings it closer to the exit.

In the ever-evolving landscape of ML, many algorithms have proven to be particularly powerful and versatile across a wide range of applications. Each algorithm has its unique strengths and is suited to different types of problems. For example, linear regression, a cornerstone of predictive modeling, serves as a quintessential example. By analyzing historical data to establish relationships between variables, it enables businesses to forecast outcomes such as real estate pricing, providing a straightforward yet powerful tool for strategic decision-making. Logistic regression, although similarly named, excels in binary classification tasks. This algorithm is used to detect spam emails, distinguishing between spam and legitimate messages by analyzing patterns in the data. Such applications highlight the algorithm's ability to transform mundane tasks into automated processes, enhancing operational efficiency.

Decision trees and their ensemble counterpart, random forests, bring a different flavor to the table. These algorithms excel in segmentation and classification, making them invaluable in credit scoring or designing targeted campaigns. By partitioning data into meaningful subsets, they enable businesses to tailor their strategies with high accuracy, whether it's crafting targeted marketing campaigns or assessing credit risk. Support vector machines (SVM)[9] are very useful in handling complex classification tasks such as image classification. SVMs can distinguish between objects in photographs with remarkable precision, illustrating the algorithm's capacity to handle high-dimensional data and uncover intricate patterns.

In the realm of recommendation systems, K-nearest neighbors (KNN) shines by leveraging the power of similarity. By analyzing the purchasing behaviors of similar users, KNN can suggest products that resonate with individual preferences, driving personalized user experiences and boosting sales. K-means clustering further enriches the toolkit by enabling customer and market segmentation through unsupervised learning, grouping customers based on buying behavior to inform strategic marketing decisions. With its probabilistic approach, the naïve Bayes model excels in text classification tasks like sentiment analysis or marketing models.[10]

Neural networks, the bedrock of deep learning, and gradient boosting machines (GBMs) illustrate the sophistication of modern ML. Neural networks power speech recognition systems, transforming audio inputs into actionable data, exemplified by virtual assistants like Siri and Alexa. GBMs, however, are adept at predictive maintenance in industrial settings, analyzing sensor data to preempt equipment failures. These algorithms underscore the transformative potential of ML, not just in automating tasks but in providing predictive insights that drive proactive decision-making. As we explore further into deep learning and generative models, the horizon of AI expands, promising even more advanced solutions to the complex challenges faced by modern enterprises.

ML, with its various approaches and the critical process of feature engineering, offers powerful tools for tackling a wide array of problems. By transforming raw data into meaningful features and iteratively refining models, these tools can unlock new levels of insight and predictive accuracy. This dynamic process not only enhances decision-making but also drives innovation and competitive advantage in an increasingly data-driven world.

Deep Learning

Deep learning is a subset of ML that uses neural networks with many layers (hence "deep") to automatically learn and transform features from raw data into higher-level representations. It differs from traditional ML because it does not rely on manual feature engineering, instead automatically discovering and optimizing features through the training process. Performing this feature transformation is one of the key strengths of deep learning, enabling it to handle complex data and tasks such as image recognition, natural language processing, and speech recognition in a way that emulates a human brain.

These models automatically extract and transform features from raw data without the need for manual feature engineering. This enables the models to identify complex patterns in data like images and videos that would be challenging and time-consuming to define manually. Deep learning networks create hierarchical layers of representations, where each layer captures increasingly abstract features of the input data. For example, in image processing, early layers might detect edges, and deeper layers recognize objects and scenes. These models benefit significantly from large datasets and high-performance computing resources. The ability to scale with data enables these models to continue improving as more data becomes available, unlike traditional ML models, which might plateau. They also allow for end-to-end learning in which the model learns to map inputs directly to outputs. This is particularly useful in complex tasks such as speech recognition and language translation, in which the entire pipeline from raw data to final prediction is optimized jointly.

Deep learning represents a paradigm shift in the realm of AI, characterized by its ability to learn from vast amounts of data and extract intricate patterns through layered neural networks. At its core, deep learning mimics the human brain's neural networks, enabling machines to perform complex tasks that were once considered the exclusive domain of human intelligence. This profound capability has sparked a revolution across various industries, driving innovations and unlocking new potentials.

Convolutional neural networks (CNNs) are perhaps the most iconic deep learning models, renowned for their prowess in image and video recognition. By leveraging convolutional layers, which scan the input data for visual features, CNNs excel at identifying objects and patterns in images. This technology is the backbone of facial recognition systems, enabling security applications that can identify individuals in real time. Moreover, in the medical field, CNNs can analyze medical images, such as X-rays and MRIs, with remarkable accuracy, identifying anomalies and early signs of diseases like cancer. For instance, CNNs have been instrumental in enhancing the accuracy of mammogram readings, allowing for earlier detection and treatment of breast cancer. This capability not only improves patient outcomes but also alleviates the burden on medical professionals, enabling them to focus on complex cases that require human judgment.

Recurrent neural networks (RNNs), and their advanced variants like long short-term memory (LSTM) networks, shine in handling sequential data, making them indispensable for tasks involving time series and natural language processing. RNNs can remember previous inputs through their internal state, enabling them to make predictions based on historical data. This capability is pivotal in applications such as language translation, where the context of previous words is crucial for accurate translation.[11] LSTMs, with their enhanced ability to capture long-term dependencies, are employed in speech recognition systems, powering virtual assistants like Amazon's Alexa and Google Assistant, which can understand and respond to spoken language with remarkable accuracy.

In the automotive industry, deep learning is the driving force behind the development of autonomous vehicles. These self-driving cars rely on a combination of CNNs and RNNs to navigate and interpret their surroundings.[12] CNNs process the visual data captured by the vehicle's cameras, identifying objects such as pedestrians, traffic signs, and other vehicles. Meanwhile, RNNs handle the temporal aspects of driving, predicting the movement of objects over time to make real-time decisions. Companies like Tesla and Waymo have leveraged these deep learning models to create vehicles that can autonomously drive in various conditions, promising a future with fewer accidents and more efficient transportation systems.

Generative adversarial networks (GANs) represent another frontier in deep learning, comprising two neural networks – the generator and the discriminator – that work in tandem to create realistic synthetic data. GANs have revolutionized fields such as computer graphics, enabling the creation of highly realistic images and animations. In the realm of entertainment, GANs are used to generate lifelike avatars and special effects that blur the line between reality and fiction. Furthermore, in the health care sector, GANs are applied to augment training data for medical imaging, where the availability of diverse and high-quality datasets is often a bottleneck. Moreover, GANs are employed in music production, where they can compose original pieces that emulate the style of renowned composers, pushing the boundaries of creative expression.

Transformers, a relatively recent innovation, have dramatically advanced the capabilities of natural language processing. Unlike RNNs, transformers process entire sequences of data simultaneously, enabling them to handle long-range dependencies with ease. This architecture underpins models like bidirectional encoder representations from transformers (BERT)[13] and generative pre-trained transformers (GPT),[14] which have set new benchmarks in tasks such as language understanding, text generation, and conversational AI. These models can comprehend context, sentiment, and nuance, making them invaluable for applications like virtual assistants and chatbots. For instance, transformers power chatbots that can engage in nuanced, context-aware conversations, enhancing customer service experiences across various industries. Transformers also power conversational AI systems that provide customer support, offering personalized and context-aware responses. This not only enhances customer satisfaction but also enables businesses to operate more efficiently by automating routine inquiries.

These deep learning models and algorithms are not merely incremental improvements but represent fundamental leaps in AI capability. They enable machines to perceive, understand, and generate content in ways that closely resemble human cognition, transforming how businesses operate and innovate. As deep learning continues to evolve, its applications will undoubtedly expand, driving further advancements and creating new opportunities in fields ranging from health care and finance to entertainment and beyond. The journey into the depths of AI is just beginning, promising a future where the boundary between human and machine intelligence becomes increasingly blurred.

Generative AI

Generative AI refers to a subset of AI focused on creating new data instances that resemble the training data. Unlike traditional discriminative models, which learn to predict labels for given inputs, generative models learn the underlying distribution of the data and can generate new, similar data points. This capability of data creation is used in various applications, such as image synthesis, text generation, and music composition.

Generative AI models like GANs and variational autoencoders (VAEs) create new data instances that resemble the training data. This enables the generation of realistic images, text, and audio, which can be used to augment datasets or create entirely new content. By learning the distribution of normal data, generative models can identify anomalies or outliers. This is particularly useful in applications such as fraud detection, where the model can flag unusual patterns that deviate from the learned normal behavior.

Generative AI represents a profound shift from merely analyzing data to creating new content. This capability stems from advanced models like GANs and transformer-based architectures, which have the remarkable ability to generate text, images, and even entire videos that are indistinguishable from those created by humans. Generative AI is not just a technological marvel; it is a powerful tool that is driving innovation across various industries by automating creativity and augmenting human capabilities. These models can generate creative outputs such as artworks, music compositions, and written stories. They learn underlying patterns and styles from the training data, enabling them to produce novel, high-quality creative content. Generative AI can also create personalized content and recommendations based on user preferences and behavior. This includes generating personalized marketing messages, product recommendations, and adaptive learning materials.

Generative AI captured mainstream attention in late 2022 with the launch of ChatGPT,[15] a chatbot capable of remarkably human-like interactions. Developed by OpenAI, ChatGPT quickly became a global sensation, captivating the public's imagination. Alongside ChatGPT, OpenAI's DALL·E 2 tool showcased another facet of generative AI's potential by creating striking images from textual descriptions.

Beyond chatbots, generative AI has become a catalyst for creativity and efficiency. Hollywood studios, for example, are leveraging GANs to produce realistic special effects and digital characters that blend seamlessly with live-action footage.[16] These AI-generated elements can significantly reduce production costs and time, enabling filmmakers to bring their most ambitious visions to life without the limitations of traditional computer-generated imagery techniques. Moreover, in the gaming industry, generative models are used to create expansive, procedurally generated worlds that offer unique experiences for each player. This not only enhances gameplay but also reduces the manual effort required to design complex environments.

The fashion industry is another arena where generative AI is making waves. Designers are using AI models to create new clothing designs by learning from vast collections of existing fashion items.[17] These AI-generated designs can suggest novel combinations of patterns, colors, and styles that might not have been considered otherwise. Brands can also use generative AI to personalize fashion recommendations for customers, tailoring suggestions to individual tastes and preferences. This level of customization not only enhances customer satisfaction but also drives sales by providing shoppers with products that resonate with their unique sense of style.

In the world of art, generative AI is pushing the boundaries of creative expression. AI-generated artwork, created by models trained on vast datasets of paintings and illustrations, is gaining recognition and value in the art market. For instance, the portrait "Edmond de Belamy," created by a GAN, was auctioned at Christie's for a staggering $432 500,[18] highlighting the potential of AI as a legitimate tool for artistic creation. This convergence of technology and art opens new avenues for collaboration, where human artists and AI systems cocreate pieces that challenge our perceptions and expand the horizons of artistic possibility.

Marketing and content creation are also being revolutionized by generative AI. AI-powered tools can generate high-quality written content, such as articles, social media posts, and marketing copy, at scale. This automation enables businesses to maintain a consistent and engaging online presence without the constant need for human writers. Additionally, AI-generated content can be tailored to specific audiences, optimizing engagement and conversion rates. In advertising, generative AI can create personalized ad campaigns that dynamically adjust to user behavior and preferences, delivering highly targeted and effective marketing messages.

The rapid rise of generative AI brings with it significant and evolving risks. Already, malicious actors have exploited this technology to create "deep fakes," counterfeit products, and intricate scams. Tools like ChatGPT, which are trained on vast amounts of public data, are not designed to comply with regulations such as General Data Protection Regulation or copyright laws. Enterprises must therefore be vigilant about how they use these platforms. Key oversight risks to monitor include a lack of transparency because generative AI models can be unpredictable and even their creators sometimes struggle to fully explain their workings.[19] Accuracy is another concern because these systems can generate inaccurate or fabricated outputs, making it crucial to verify all outputs for correctness and relevance before use. Additionally, there is the issue of bias; companies need to implement policies to detect and address biased outputs, ensuring compliance with both legal and internal standards.

Gartner advises enterprises to consider critical questions such as defining responsible use of generative AI, ensuring compliance, managing consent, and protecting IP. Organizations must also consider the broader impacts on trust and how new economic models might be structured to fairly compensate content creators. By staying informed and proactive, organizations can navigate the complex landscape of generative AI, balancing innovation with responsible and ethical use.

Generative AI is not just a technological advancement; it is a transformative force that is reshaping industries by automating and augmenting the creative process. Its applications span a wide array of fields, from entertainment and fashion to art and marketing, each benefiting from the unprecedented capabilities of AI-generated content. As we continue to explore and harness the potential of generative AI, it is clear that this technology will play a pivotal role in driving innovation and redefining the boundaries of human creativity.

Gartner predicts that generative AI will evolve into a general-purpose technology with an impact comparable to that of the steam engine, electricity, and the internet. Although the initial hype might eventually subside as the practicalities of implementation become apparent, the transformative potential of generative AI will only increase. As individuals and enterprises explore more innovative applications, this technology is poised to revolutionize daily work and life, driving unprecedented levels of efficiency and creativity.[20] The journey of generative AI is only beginning, and its impact on our world is poised to be profound and far-reaching (see Figure 4.2 for the time line of key milestones in AI).

Machine Intelligence and Biologically Inspired Models

Machine intelligence and biologically inspired models represent the next frontier in AI, aiming to bridge the gap between artificial systems and the complex, adaptive intelligence observed in nature. These models draw inspiration from the principles and mechanisms of biological systems, leveraging the nuanced understanding of how natural intelligence operates to develop more sophisticated and resilient AI systems.

Machine intelligence extends beyond traditional ML by incorporating elements of reasoning, learning, and perception that are closer to human cognition. Biologically inspired models, such as neural networks modeled after the human brain, evolutionary algorithms that mimic natural selection, and swarm intelligence based on the collective behavior of social insects, embody this approach.

Building Intelligence

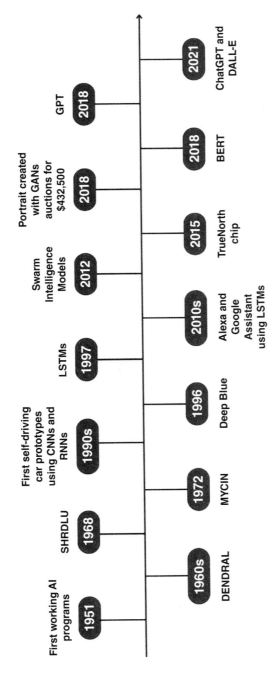

FIGURE 4.2 Time line of key milestones in artificial intelligence: This time line highlights significant achievements in AI from the first working AI programs in 1951 to the advent of advanced models like ChatGPT and DALL-E in 2021. Key developments include the creation of expert systems in the 1960s and 1970s, breakthroughs in deep learning and neural networks in the 1990s, and the rise of generative models and advanced AI applications in the 2010s and 2020s.

These models are designed to adapt, learn, and evolve over time, providing robust solutions to complex, dynamic problems.

The main defining characteristics of machine intelligence are found in its architectural inspiration, the brain. These models use artificial neurons and layers to process information, which enables the handling of complex tasks like image and speech recognition to mimic how the human brain would handle them. These models learn to make decisions by receiving rewards or punishments, similar to how animals learn from interactions with their environment. This approach is used in robotics and game playing. These models mimic biological learning processes to continuously improve their performance by adapting to new data and experiences, making them robust to changing environments.

In the previous sections, we talked about neural networks and RL. Neural networks are perhaps the most well-known biologically inspired models since their structures mimic the human brain's interconnected neurons.

RL takes inspiration from behavioral psychology, where agents learn to make decisions by receiving rewards or penalties. For example, RL has been instrumental in training AI to play complex games like Go and StarCraft II,[21] where the AI learns strategies and adapts to human players. In industrial automation, RL optimizes robotic processes by learning the most efficient ways to perform tasks, reducing operational costs and increasing productivity.

Inspired by the process of natural selection, evolutionary algorithms evolve solutions to optimization and search problems over successive iterations. They start with a population of potential solutions and use genetic operators like mutation, crossover, and selection to evolve better solutions. In engineering, evolutionary algorithms are used for design optimization, such as improving aerodynamic properties of aircraft components or optimizing financial portfolios for maximum returns while managing risk. These algorithms adapt over time, finding innovative solutions that might not be apparent through traditional methods.

Swarm intelligence models, such as ant colony optimization and particle swarm optimization, draw from the collective behavior of social organisms like ants, bees, and birds. These models are particularly effective in solving complex optimization problems and coordinating multiple agents. For example, ant colony optimization is used in logistics and supply chain management to optimize routing and scheduling, mirroring how ants find the shortest path to food sources.[22] Particle swarm optimization has applications in robotics for path planning and obstacle avoidance, enabling robots to navigate efficiently in dynamic environments.[23]

Neuromorphic computing aims to emulate the neurobiological architectures present in the human brain. This approach involves designing hardware that mimics neural structures, allowing for more efficient and powerful processing capabilities. IBM's TrueNorth chip is an example, designed to perform brain-like computations with minimal energy consumption.[24] Applications of neuromorphic computing are found in real-time data processing for edge computing devices, enabling advanced capabilities in Internet of Things systems, such as real-time anomaly detection in industrial equipment.

Machine intelligence and biologically inspired models represent a significant leap forward in the field of AI, bringing us closer to creating systems that can think, learn, and adapt in ways similar to natural intelligence. By drawing from the principles of biology, these models offer robust, flexible, and innovative solutions across various domains, from health care and engineering to logistics and robotics. As we continue to explore and refine these models, the potential for transformative impacts on technology and society is immense, heralding a future when AI not only augments human capabilities but also inspires new ways of thinking and solving problems. The journey into machine intelligence is just beginning, and its trajectory promises to reshape our world in profound and unexpected ways.

To conclude, we have embarked on a journey through the landscapes of AI, exploring the transformative power of ML, deep learning, and generative AI, as well as the profound insights offered by biologically inspired models. Each of these technologies, with their unique capabilities and applications, represents a significant leap forward in our quest to replicate and enhance human intelligence.

In the chapters that follow, we will delve deeper into real-world examples of these models in practice, illustrating how they are being leveraged across various industries to solve complex problems and drive innovation. Additionally, we will explore the fascinating intersection of AI and behavioral economics, revealing how the combination of these fields can help identify the most crucial signals and uncover real value. By understanding human behavior and decision-making processes, we can fine-tune AI models to better predict outcomes, personalize experiences, and ultimately create more impactful solutions.

As we continue this exploration, the goal is not just to appreciate the technological advancements but also to harness them in ways that are strategically aligned with organizational goals and ethical considerations. Together, we will navigate this exciting frontier, unlocking new potentials and charting a course toward a future when AI and human ingenuity coexist harmoniously to create a better world.

CHAPTER 5

Real-World Impact
Harnessing AI and ML for Practical Solutions

"AI will be the most transformative technology of the twenty-first century. It will affect every industry and aspect of our lives."

– Jensen Huang

In this chapter, we bring together the theories and concepts from our previous discussions on artificial intelligence (AI), machine learning (ML) algorithms, and behavioral economics into actionable insights. We've delved into definitions, mechanisms, and intricacies. Now, let's see how these powerful tools can address complex challenges and drive innovation across various industries. This chapter aims to bridge the gap between theoretical understanding and practical application, showing how AI and ML can transform abstract concepts into real-world solutions.

Through a series of real-world examples and case studies, we will discover how predictive models, enriched with behavioral insights, can uncover hidden patterns in data, enhance decision-making, and create substantial value. From optimizing supply chains and improving customer experiences to predicting consumer preferences, the stories in this chapter will highlight the transformative potential of AI and ML. We will see how the ideas we've explored come to life, offering a clear road map for leveraging these technologies to achieve strategic objectives. This is where everything comes together, illustrating how AI and ML can turn data into a powerful driver of growth and innovation.

Unleashing the Full Potential of AI: Beyond the Hype

In an era dominated by headlines touting the extraordinary capabilities of generative AI, it is crucial for executives to appreciate the broader spectrum of opportunities that advanced predictive models present. Although generative AI captivates with its ability to create content and simulate human-like interactions, the true power of AI extends far beyond these applications. This chapter aims to illuminate the diverse applications of AI and ML models, showcasing how they can be strategically leveraged to drive efficiencies, generate revenue, and de-risk critical decisions.

For executives, understanding these varied applications is vital. The hype around generative AI can sometimes overshadow the profound impact that more traditional AI models can have when thoughtfully implemented. By integrating predictive models with behavioral insights, organizations can unlock unprecedented efficiencies and revenue-generation opportunities. This combination allows for a deeper understanding of customer behavior, enabling highly targeted interventions and strategies that drive meaningful engagement and results.

Moreover, the ability to experiment on a smaller scale before scaling interventions or initiatives is a game changer. This approach not only mitigates risk but also uncovers novel strategies and solutions that might not have been considered otherwise. Through experimentation, organizations can refine their approaches, ensuring that only the most effective and impactful strategies are implemented at scale. This iterative process of testing and learning is essential for building resilience and adaptability in today's fast-paced business environment.

As we delve into specific examples of AI applications, we will explore how these models, when combined with a nuanced understanding of human behavior, can transform operations across various sectors. From optimizing marketing spend and personalizing customer experiences to enhancing demand forecasting and inventory management, the potential is vast. By broadening our perspective and embracing the full range of AI capabilities, we can drive sustainable growth and innovation, positioning our organizations for long-term success.

Rethinking Segmentation: Beyond Demographics and Life Stages

The first time I learned about life-stage-based segmentation was a few years ago while working with a credit union, and I must say it was a strange concept. Not because it wasn't useful, but because it was extremely limited. I briefly talked about segmentation and clustering models in Chapter 4 when discussing ML models. These models are unsupervised models that can be built using a rule-based approach with a handful of variables or can capture hidden patterns in the data using hundreds of variables and features. Life-stage segmentation used in most banks and financial institutions only uses age and a handful of available demographics to divide a market into groups based on the different stages of life people go through, such as adolescence, adulthood, and retirement. These stages supposedly have different financial implications due to job and family status, as well as overall lifestyle.

Being a behavioral economist, I am trained to consider and uncover a multitude of factors that affect human behaviors, decisions, and outcomes. An economist considers the incentives, motivations, social influences, and more when it comes to understanding why an individual decides to open a bank account further away from where they live, how they allocate their savings, or how they invest their money using different financial instruments. For instance, if your mother, like mine, constantly reminds you to save as much as possible, the odds are you will. Similarly, if your friends are actively investing, you might ask about the stocks they are investing in and the returns they see (and, of course, be exposed to some level of the biases and heuristics guiding their decisions).

This broader perspective led me to realize the limitations of life-stage-based segmentation. It fails to capture the complex and dynamic nature of human behavior, which is influenced by myriad factors beyond just the stage of life.

Have you seen the movie *Dumb Money*?[1] It brilliantly illustrates a fascinating phenomenon in which investment decisions were driven not by age or gender but by peer effects and social networks. The film chronicles the GameStop short squeeze of January 2021, highlighting how individual investors, fueled by discussions on Reddit and other social platforms, collectively drove up the stock price against hedge funds' expectations. This real-world example underscores the immense power of social influence and community-driven decisions in the financial market.

This is precisely why the concept of a segmentation model based solely on one dimension, such as age, seemed so limited to me. Especially in the realm of financial decisions, where individuals often do not act rationally, as Richard Thaler's Econs would suggest, reducing people to a single characteristic misses the complexity of human behavior.

To their credit, the executives at this credit union recognized that this simplistic approach was insufficient, which is why we were having this conversation. They wanted to understand the deeper drivers of behavior and engagement among their members to serve them better and present the right product at the right time. Instead of calling their members to offer them a random product, they aimed to delight them with an offer that was precisely tailored to their needs. They understood that they needed more than just the age of their members to guide these decisions effectively.

So we embarked on a journey to uncover every possible behavioral signal hidden in the data. We aggregated data on various aspects of behavior, including savings, wealth management, insurance, loans, mortgages, payroll, and credit cards. We didn't just explore the raw data; we engineered over 300 features to capture factors such as changes in saving behavior over the last three months and how these changes compared with the rest of the members or with their own historical behavior over the past few years. Our goal was to detect significant life events and changes, such as going to university or having a baby, as well as subtle shifts in preferences that aligned with macroeconomic conditions.

We wanted to identify what had changed this year compared to last year, which might indicate that a different financial product would be more appropriate for their current needs. By communicating these insights proactively, we aimed to reach the customer before they started shopping around at different banks and financial institutions.

We used several different unsupervised ML models, such as backward elimination segmentation models, where we made no assumptions about the number of member segments or their size and defining characteristics ex ante. We fed all the features into the model, enabling it to explore every possible permutation and eliminate features one by one to arrive at the most stable segments. These segments grouped members based on their similarities across a variety of behavioral factors. We contrasted this approach with more classical segmentation methods, such as simple K-means and hierarchical models, which allowed for assumptions about the number of segments and the specific behavioral features considered.

Six unique segments were identified as a result of this exercise, grouping members based on their engagement levels, risk attitudes, liquidity, borrowing, and investing behaviors, among other factors. Interestingly, we found no correlation between the segments and age or gender. We had individuals in their 30s and their 70s within the same segment, indicating that life stages are not the only important factor in determining the financial needs of customers. If we hope to serve members and customers with products and services that truly enhance their financial well-being, we must go beyond mere demographics.

By combining these advanced analytical techniques with our deep understanding of behavioral economics, we were able to develop a segmentation model that went far beyond the traditional life stage approach. This enabled us to provide highly personalized and relevant financial product recommendations, enhancing customer satisfaction and loyalty while driving significant business value for the credit union.

Uncovering Unexpected Customer Patterns

The same challenge often arises in retail and even supply chain industries. These sectors are fundamentally customer- and consumer-centric, yet when it comes to understanding customers and their levels of engagement or satisfaction, we frequently overlook the subtleties of social influence, peer effects, biases, and heuristics, among other motivating factors. Deep behavioral segmentation is one way to systematically capture these subtleties and uncover patterns and true drivers of observed behaviors.

One of the biggest challenges I've observed in some retailers is the separation of in-store and offline behavior from online shopping behaviors. The segmentations and persona creations I've seen often rely on aggregate metrics such as the total number of items purchased, total dollars spent in a month (both online and offline), or high-level product information combined with available demographics. These are then used in cohort analysis or rule-based methods to create customer segments.

The situation in supply chain management is even more problematic. One of the critical factors affecting customer experience – whether they are businesses or end consumers – is disruptions along the supply chain. These disruptions lead to delays, missed target service levels, and failures to deliver on time and in full. Given this, it only makes sense to consider the multiple factors that influence customer experience and can lead to complaints and disputes. However, due to the operational nature of these organizations, commercial teams often build segmentation models based on just two or three operational metrics, such as total revenue or volume per customer, without considering any behavioral factors. These might include how many times customers have called customer service to inquire about a delayed shipment, how often deliveries were missed, or items were returned due to damage.

This problem is exacerbated by the operational model and the silos that exist within these organizations, making it challenging to access relevant information in the first place. To truly enhance customer experience, a more integrated approach is needed – one that bridges these silos and considers the full spectrum of customer behavior and engagement across all channels and touchpoints.

All that said, I love working with retailers because they usually have access to a vast amount of data, yet they often struggle to extract maximum value from it. We introduce them to the concept of minimum viable data (MVD), explaining that by identifying the right datasets, we can build essential features and variables for a high-accuracy model. Little do they know, we will request datasets they have never connected to one another before.

We integrate datasets that detail product attributes with transactional data from stores and e-commerce platforms that have detailed event-based information. We examine returns, whether they happen in-store or online. We consider how many different items a customer looked at online before making a purchase, accounting for phenomena like choice fatigue and loss aversion. We analyze how quickly a shopping session was completed, how many different colors a customer clicked on, and whether they abandoned their shopping cart only to return to it a day later.

Our goal is to build hundreds of features to capture as many different aspects of behavior and the driving factors as possible. We then feed all this data into an unstructured model like backward elimination, which doesn't require any ex ante assumptions. This comprehensive approach enables us to uncover deep insights and drive significant improvements in customer engagement and operational efficiency.

Often, very interesting patterns arise that defy conventional assumptions typically tied to age, gender, or socioeconomic characteristics. For instance, sometimes the most valuable customers, in terms of spending, are actually discount shoppers. These customers not only wait for discounts but are frequent visitors to stores and websites to avoid missing out on good deals – illustrating loss aversion once again.

Moreover, the availability of complementary products significantly increases the chances of these discount shoppers spending more. Interestingly, these customers are not necessarily more price sensitive than others. They simply love a good deal or are more risk averse. This highlights the need to move beyond traditional segmentation and assumptions to truly understand and cater to customer behaviors.

Another example is customers who return items. Retailers strive to minimize returns or make the experience as seamless as possible for various reasons: the cost of returns, inventory optimization, and, more important, minimizing bad customer experiences that might lead to attrition and churn. However, there are different kinds of returners. Some people systematically order more, especially online, to try different colors and sizes. They purchase with the intention to return and have very distinct behavioral signatures. By contrast, other individuals spend significant time and effort comparing different items and reading reviews to make well-informed decisions and pick the right product.

Interestingly, loss aversion is at play within both of these groups. One group spends time in advance to minimize the need to (1) spend more money than necessary and wait for a refund, and (2) go to the store and wait in long queues to return items. The other group, perhaps less price sensitive, wants to ensure the fit and look of the product are perfect and prefers having a few options to compare.

When done correctly and comprehensively, descriptive models such as segmentation can shed light on many different aspects of behavior and help build hypotheses. By understanding these nuanced behaviors, retailers can create more targeted strategies that not only enhance customer satisfaction but also improve operational efficiency. These insights provide a foundation for developing predictive models and interventions that can significantly affect business outcomes, fostering a deeper connection with customers and driving sustained growth.

Predicting Intent and Mapping Customer Journeys

A topic that goes hand in hand with clustering and behavioral segmentation, especially for retailers and brands, is predicting intent. Behavioral segmentation helps develop a deep understanding of customers and consumers – their needs, preferences, and how they compare with the rest of the population. However, it is not very actionable because it doesn't answer the crucial question of "so what?" This is where predicting intent and mapping customer journeys become vital.

Another important topic for retailers and brands is mapping customer journeys. This is less trivial than it sounds. In one of the large grocers we worked with just before and halfway through the pandemic, the teams had spent months collecting data to understand the most common journeys customers take and the correlations between these journeys. For example, they analyzed how often a routine shopping trip is combined with an emergency one, such as having guests with little notice, when the customer only picks up a few items and has to rush out.

Customer journey mapping involves tracking and analyzing the various touchpoints a customer interacts with, from the initial awareness stage to post-purchase behavior. It requires understanding not just the sequence of interactions but also the context and motivations behind them, for instance, identifying patterns in how customers switch between online and offline channels, their decision-making processes, and the triggers that prompt specific actions.

These customer journeys help marketing and product teams define more actionable engagements and interventions when needed. By understanding the full journey, retailers can anticipate customer needs and provide timely and relevant interactions, ultimately enhancing the customer experience and driving loyalty.

Retailers and brands often wonder if it is possible to predict what customers intend to purchase. Imagine an online shopping trip when the customer spends 30 minutes adding more than a dozen items to their basket, only to abandon it afterward. Why does this happen? What was the customer looking for that they couldn't find, leading to the abandonment of the entire basket? Is this a routine shopping journey or an emergency one? It is understandable that they might be looking for something specific that was out of stock, especially in an emergency. However, if this occurs regularly during routine shopping trips, customer churn is likely to follow.

Being able to predict customer intent creates the opportunity to be proactive. For example, if a customer is looking for items that are out of stock, we can recommend substitute products proactively. If they show a preference for organic products by adding organic fruit to their basket and organic yogurt happens to be on sale, a complementarity-based recommendation model can delight them with this suggestion.

Additionally, detecting whether a customer is on a routine journey, making a special trip for an event like the Super Bowl, or handling an emergency trip can help guide the customer through a more enjoyable and efficient journey. This reduces the need for extensive searching and increases the likelihood of finding what they need. This proactive approach not only enhances the customer experience but also helps in retaining customers by addressing their needs more effectively.

Overcoming Challenges in Predicting Customer Intent

When we started working with this grocer, we faced multiple challenges. Before the pandemic, more than 80% of their customers shopped in stores, whereas in the first few months of the pandemic, over 95% of their customers had to shop for groceries online. Point of sale (POS) and e-commerce data sources were not connected effectively to leverage customers' past purchase history. Most of the recommendation and segmentation models were built based on online customers' activities and socioeconomic demographics, which were very different from the majority of the customers who shopped in person. Additionally, we observed that most online sessions did not end in a purchase, and the rate of cart abandonment was very high. All these challenges underscored the importance of being able to predict intent and assist and engage customers at the right moment with the right product.

To address this, we conducted a thorough behavioral segmentation and used a transformer-based neural network (NN) architecture to maximize the signals that would help us predict risk: the risk of a customer abandoning a shopping session. By projecting the outcomes onto a lower-dimensional model such as an XGBoost, we could predict the reasons behind the abandonment. For example, we could predict that a customer would abandon a session because some products they intended to order were out of stock. Alternatively, we could predict that increasing the diversity of the basket through complementarity-based recommendations would increase their purchase probability, or that price did not significantly affect most customers, indicating they were not primarily bargain hunting.

Notice that we employ three distinct types of ML algorithms to predict the risk of customers abandoning their carts. First, we use an unsupervised segmentation model to identify patterns and group similar customer behaviors. Next, we leverage a deep learning model, specifically a transformer-based NN, which uses multiple layers of embeddings to learn product attributes and more nuanced details. Finally, we implement a tree-based ML algorithm, the XGBoost model, which offers a higher degree of interpretability compared to NNs. This enables us to pinpoint the most significant factors influencing the predicted outcomes.

The outcomes from these models, in combination with the segmentation model, enabled our client to design interventions that increased the probability of purchase and lowered the cart abandonment rate.

By integrating advanced predictive models with comprehensive behavioral segmentation, we were able to deliver actionable insights that transformed the customer experience. For instance, when we identified that a customer might abandon their cart due to out-of-stock items, we proactively recommended alternative products. Similarly, recognizing customers who valued a diverse basket enabled us to suggest complementary items, enhancing their shopping experience and increasing the likelihood of completing a purchase.

This approach also revealed that price sensitivity was not a major factor for most customers, enabling the grocer to focus on other aspects of value, such as convenience and product variety. By understanding the specific needs and behaviors of different customer segments, we could tailor engagement strategies that resonated more effectively with each group.

The combination of accurate intent prediction and targeted interventions not only improves purchase rates but also fosters greater customer loyalty and satisfaction. It demonstrates that a deep, data-driven understanding of customer behavior can drive significant improvements in both operational efficiency and customer experience. This holistic approach to customer engagement is crucial for any retailer aiming to thrive in a rapidly evolving market landscape.

Predicting and Managing Returns

Another example of using a similar NN and transformer-based architecture is predicting returns. Many retailers deal with return rates ranging from 28 to 45%, especially on e-commerce purchases. According to the National Retail Federation, returns accounted for $428 billion in lost sales for US retailers in 2020.[2] Beyond the significant revenue impact, returns also incur additional logistics costs, such as restocking and managing excess inventory. Moreover, frequent returns can distort demand forecasting, affect inventory management, and ultimately erode profit margins. Efficiently addressing returns is crucial for maintaining profitability and operational efficiency in retail.

In 2019, we partnered with a leading department store with banners in the United States and Canada to tackle this issue. The question was clear: "Can we predict who is going to return what item before the item is even purchased in an online shopping session?" The short answer is yes. However, the more important question is what the retailer will do with the predictions from the model. I remember asking this question to our stakeholder, and her answer surprised me: she told me that they would offer a 10–15% discount on items with more than a 75% probability of return and make it a final sale.

On the surface, this is a very logical answer. If the shopper accepts the discount, it will stop a return, saving the cost of the return and the hassle of managing excess inventory and redistributing it to the right stores or distribution centers. So, you might wonder why I was surprised. Let me explain.

I was surprised for two reasons. First, despite the model's ability to predict returns with a high level of granularity and personalization – at the individual shopper, basket, and item level – the proposed solution was a one-size-fits-all approach. This seemed counterintuitive given the personalized nature of our predictions. Each customer's behavior and reasons for returning items are unique, and a blanket discount strategy might not address the underlying issues effectively.

Second, by implementing such a generic solution, the retailer could miss out on valuable insights into customer behavior. For instance, some returns might be due to poor product descriptions, sizing issues, or even seasonal trends. Understanding these nuances can help retailers not only reduce returns but also improve their overall product offerings and customer experience. Instead of a broad discount policy, targeted interventions based on specific return reasons could provide more meaningful and long-term solutions.

Although our stakeholders understood the concept of predictive models and probabilistic outcomes, they assumed the model was always "right" when predicting returns with over 75% probability. Well, that's not exactly the case. There are always false positives and false negatives, which is why data scientists usually report various metrics such as model accuracy, precision, and recall, among others.[3]

To make this a bit clearer, let's imagine a scenario in which an individual was not planning to return an item, and the model incorrectly labeled the item as a "return" (i.e. a false positive), thereby offering a discount to a customer who was going to purchase the product at full price anyway. In this case, not only do we fail to save the cost of the return because the customer was not going to return it, but we also lose margin due to the discount.

This scenario illustrates the trade-off between different key performance indicators (KPIs) and model performance metrics and the real implications of generalized interventions. We conducted a thorough analysis for our client that quantified this trade-off based on the performance metrics of the model and the cost of returns and other business KPIs. We showed that if they were not careful, the intervention could lead to losing $750 000. But more important, through simulations, we demonstrated how personalized interventions based on the same information used to predict the probability of return could save them $6 million. To illustrate this, we showed how the outcomes from the predictive model could be used as inputs to the behavioral segmentation model. This model could then separate different types of customers and help identify more personalized interventions, such as lowering the free shipping threshold and offering 5% to 10% to 15% discounts to different groups to increase margins.

By using predictive analytics and personalized strategies, retailers can not only minimize the costs associated with returns but also enhance overall customer satisfaction and loyalty. This holistic approach ensures that interventions are not just reactive but are tailored to meet the unique needs of each customer segment, leading to sustainable growth and profitability. Leveraging advanced predictive models and behavioral segmentation can transform how retailers engage with customers, optimize inventory management, and enhance overall profitability. By understanding customer intent and personalizing interventions, retailers can significantly reduce cart abandonment rates and return rates, leading to a much more efficient and customer-centric operation.

This comprehensive strategy fosters a deeper connection with customers and drives long-term success in an increasingly competitive market.

The Power and Nuances of Recommendation Models

Recommendation models were popularized by Amazon and other shopping apps, as well as news and social media outlets. We are all familiar with the experience of seeing a few suggested items pop up after completing a shopping session on Amazon, or encountering "suggested" posts while scrolling through social media. I remember a friend of mine who was getting married asked me to search something on my phone, and for the next three months, I kept getting recommendations for all things wedding-related on my social media accounts.

Have you ever wondered how these models work? What kind of data do they use to recommend products that are actually relevant instead of being annoying or feeling intrusive? However, what is the success rate of these recommendation engines when many of us find them annoying or simply ignore them? Why do they work in some contexts and not in others? How can they be used to nudge behavior effectively, uncover useful products or offers that delight us, or help us discover items we wouldn't have found otherwise?

Understanding the mechanics and data behind recommendation models can shed light on their varying success rates. These models typically use vast amounts of data, including past purchase history, browsing behavior, and demographic information, to predict what products or content might interest a user. The goal is to make personalized recommendations that feel natural and helpful rather than intrusive.

However, the effectiveness of these models depends on several factors. In some cases, recommendations can feel spot-on and enhance the shopping experience, and in others, they miss the mark and become a source of frustration. The context in which these recommendations are presented, the quality and relevance of the data used, and the algorithms' ability to adapt to individual preferences all play crucial roles.

When designed and implemented effectively, recommendation models have the potential to significantly enhance customer engagement and satisfaction. They can uncover products or content that users might not have found on their own, providing a sense of discovery and delight. By leveraging advanced analytics and continuously refining their algorithms, businesses can use recommendation models not only to boost sales but also to build stronger, more personalized connections with their customers.

There are many kinds of recommendation models, but in essence, they all aim to achieve the same goal: use historical patterns on similar products purchased by similar individuals to predict the next best product or offer for a customer. The concept of "similar" is crucial here because these models sift through massive amounts of data to identify two axes of similarity. First, they determine who is behaviorally similar to us and then identify which products similar individuals have purchased that align with our past purchases. This provides clues about products we haven't yet purchased but might be interested in.

To build high-accuracy precision models, several years of historical data are typically required to capture products purchased at a basket-customer level. Most brands, retailers, grocers, financial institutions, wealth management firms, hospitality, and tourism industries possess this kind of data, even if it is not perfectly cleaned or organized.

If an organization can accurately identify which products were purchased by whom over several years and has a substantial sample size (e.g. millions of transactions by thousands or millions of customers over two to five years), it is possible to build recommendation models that predict the next best products at a personalized level. One of the initial steps in building these models is applying some version of an identity resolution algorithm to consolidate customer information and resolve duplication.[4] The next crucial step is creating relevant features that capture various aspects of the products. The more detailed and granular these descriptions and definitions, the better, because NN models will have a richer dataset to learn from. Merely feeding stock-keeping units (SKUs) and product codes to the model will capture only the relationship with characteristics of individuals, their preferences, and the products to some extent. Additionally, many retailers recycle product and SKU codes for different, unrelated products, complicating the process.

This all sounds good in theory, but life as a data scientist is not always easy. What if we don't have a lot of data? Or what if it is challenging or impossible to connect basket- and product-level data to the individual who made the purchase? Or what if we have a small number of transactions on a few products (think about a bank with many customers but only a few hundred unique products)? Can we still build a recommendation model that suggests the next best product?

The answer lies in creatively leveraging the data we do have. For instance, even with limited data, clustering techniques and behavioral segmentation can help identify patterns and similarities among customers. Collaborative filtering, content-based filtering, or hybrid approaches can be employed to make the most of the available data. Additionally, leveraging external data sources, such as social media interactions or publicly available demographic information, can enrich the dataset and enhance the model's predictive power.

Broadening Horizons: Beyond Category Killers

Let me give you two nontrivial examples. A few years ago, we were working with a wealth management firm in New York City. They managed close to $500 billion in assets and had 55 regional managers who worked with over 80 000 financial advisors across the United States. Their product portfolio included highly rated offerings such as fixed income or tax-free income investments, often praised on platforms like Morningstar.

When we met with the chief transformation officer and other executives, they kept referring to their "category killers." These were their best-selling products, known for their stellar market performance and widespread recognition. Regional managers frequently pushed these products because they performed exceptionally well, practically guaranteeing bonuses. However, this created a significant challenge: limited product diversification. Regional managers focused on selling a handful of category killers, leading to a loss in potential market share.

Why would they venture beyond these high-performing products when their bonuses were ensured by sticking to them? This mindset was precisely why the firm wanted to tap into their data to identify other products their clients might be interested in. They aimed to expand product offerings for current clients beyond category killers, predict financial advisors' product preferences, and drive revenue growth through healthy product diversification.

By leveraging advanced data analytics and recommendation models, the firm sought to understand the nuanced preferences of their financial advisors and clients better. This approach would not only diversify the products being offered but also help maintain a competitive edge in the market.

This scenario exemplifies a situation where, despite having tens of thousands of customers with numerous transactions, most purchases were concentrated on two to three different products. To address this, we needed to tap into a wealth of external data to uncover the financial advisors' preferences, account for the macroeconomic environment, and understand the true performance of the remaining products compared to the broader offerings of their competitors.

We developed a product recommendation module using dynamic recursive neural networks (RNNs), which are well suited for processing sequential data and capturing temporal dependencies. RNNs excel at understanding patterns over time, making them highly effective in predicting the dynamic, time-based behaviors of financial advisors. These networks can apply reasoning modeled after the human brain, "learning" just as a human would, but with the capacity to process and account for far more data.

The module predicts a dynamic score for potential and existing advisors based on their likelihood of buying a product, identifies the top five products or asset classes the advisor is most likely to buy, and flags potential high-yield relationships. We combined this model with a deep behavioral segmentation model to identify high-risk and high-potential advisors and map their journeys.

To increase stakeholder confidence, we validated the predicted top five products for each advisor against historical trends. This benchmarking process helped increase their conviction and trust in the recommendations from the predictive engine, facilitating healthy product diversification and identifying upsell and cross-sell opportunities. Through numerous simulations, we demonstrated that acting on these recommendations could result in an estimated revenue increase of over $2 billion in one quarter.

This comprehensive approach ensured that the wealth management firm could move beyond its reliance on "category killers" and tap into a broader range of products, ultimately driving revenue growth and maintaining a competitive edge in the market.

Enhancing In-Store Experience with Recommendation Models

The second nontrivial example comes from the grocery world. Many retailers and grocers have access to extensive transactional data, but unless there is a robust loyalty program that customers consistently use, it becomes challenging to connect customer data over time to the transactional and POS data. This disconnection makes it difficult to identify similar individuals with similar purchase behaviors, which is crucial for extracting signals about customer product preferences.

This challenge became particularly intriguing when one of the largest supermarket chains in the UK, with revenue over £18 billion, asked us to enhance the in-café experience through relevant product recommendations at the point of purchase. The goal was to drive repeat visits, increase customer loyalty, and boost basket size and value.

To tackle this challenge, we built a basket-based recommendation model using NN architectures. We engineered features to teach this NN various characteristics of the products instead of merely using product codes or SKUs. This enabled the model to detect patterns about the best products that are often bought together by analyzing millions of transactions. We took it a step further by incorporating insights from behavioral economics, identifying anchor products in each basket – those whose presence consistently increased the probability of purchasing certain other products.

To test whether this recommendation system could increase conversion rates and basket size, we decided to conduct a proper experiment. This grocer had about 400 grocery stores across the UK. We chose 206 of the stores and integrated the recommendation module into the self-order kiosks, observing customer interactions for two months. We integrated these recommendations at a particular moment in the customer journey: Imagine yourself at the entrance of the café, selecting a few items from an easy-to-navigate menu and adding them to your basket. When you press "proceed to checkout," at that exact moment, a window pops up with three recommendations, wondering if you forgot anything – like ordering two kids' meals but no drink for yourself.

The idea was that if we could change someone's mind and influence their decision at one of the hardest moments, then we could always move the recommendation to earlier in the journey.

The basket-based recommendation model, combined with identifying the anchor products in the baskets, achieved an average conversion rate of 5% for recommendations across all cafés. Some cafés experienced conversion rates of 8–9% compared to our control group. For orders containing recommended items, recommendations added 20–40% of the basket's value. We optimized this recommendation model for basket value, considering variables like time of day, day of the week, store volume, geography, item value, probability of purchase, and margin.

These examples illustrate the transformative potential of personalized recommendation models in various industries. Whether it's diversifying product offerings in wealth management or increasing basket size and value in grocery stores, recommendation models can drive substantial business outcomes.

However, the key to success lies in understanding the nuances of customer behavior and tailoring the models to fit specific contexts. Incorporating external data, applying behavioral economics insights, and conducting rigorous experiments ensure that the recommendations are not only accurate but also actionable. This holistic approach enables organizations to move beyond basic product suggestions to delivering personalized experiences that truly resonate with their customers. In turn, this fosters loyalty, drives repeat business, and ultimately leads to sustainable growth and profitability.

Leveraging Propensity Models for Targeted Campaigns

Most customer-centric industries find it challenging to maximize the insights from their data to create targeted campaigns and offerings for their customers. In the previous section, we discussed recommendation models that identify the best product for a given customer. Activating these recommendation engines with high levels of personalization and granularity usually requires effective digital channels to reach customers. But what if a bank wants to reach out to its customers through call centers to inform them about a new financial product or campaign? Or imagine a bank or retailer experiencing customer churn that wants to create targeted offers to reduce churn and increase engagement and loyalty. How can they create targeted marketing campaigns and increase the relevancy of offers to help reduce churn and enhance the conversion rates of call center and branch teams?

To solve these problems, we need a different kind of model that effectively helps us with the reverse of a recommendation engine. In other words, we need to figure out who the best customers are for a given product, offer, or campaign. This is where propensity models come into play.

Propensity models are a very useful class of ML algorithms often used to answer these types of questions. Typically built using ML models such as XGBoosts, these models estimate the propensity (or probability) of an individual to purchase a product

Real-World Impact

or engage with a particular offer. XGBoost, short for eXtreme Gradient Boosting, is an advanced implementation of gradient boosting decision trees designed for speed and performance. It is highly efficient, scalable, and provides state-of-the-art results for many predictive modeling tasks. XGBoost's ability to handle missing data, prevent overfitting with regularization techniques, and leverage parallel processing makes it ideal for analyzing large-scale datasets in retail and finance sectors. This enables businesses to identify and target the right customers effectively.

By leveraging propensity models, banks can determine which customers are most likely to be interested in new financial products and tailor their outreach accordingly. Similarly, retailers can identify at-risk customers and create targeted offers to reengage them and reduce churn. This approach not only improves the efficiency of marketing campaigns but also enhances the customer experience by providing relevant and timely offers.

Remember the credit union I highlighted when discussing segmentation models? After that comprehensive exercise, we were all asking, "Now what? How can we take action on the insights we've learned, especially because we have identified a group of members with a high probability of churn and reduced engagement?" In discussions with executives, we learned they ran regular campaigns through their call centers, for example, for "term deposit" products – fixed-term investments offering a guaranteed interest rate for a specified period. Every couple of months, the call center agents would call about 10 000 members at random and offer them a new term deposit product at a particular interest rate. Their conversion rate was about 3%. We knew we could improve on this randomness by leveraging the thorough behavioral segmentation analysis and the many behavioral features we had created for that model.

So, we built a series of propensity models that would score the members based on their likelihood of purchasing a financial product such as term deposits. As always, we set out to run proper experiments to test the performance of the model and the proposed interventions in practice. These models provided a list of specific customers with a high propensity to uptake a particular product. Using ML and behavioral economics, the AI campaign module extracts signals and identifies the drivers behind customer behavior.

We ran several experiments with similar designs: we provided lists containing 3000–4000 members (instead of 10 000) with very simple talking points that were specific to the member. These talking points were developed using the reason codes extracted from the ML model as the most important factors contributing to a high probability of uptake for this particular financial product for the specific member. The spring campaign generated $30.2 million in total over eight weeks. The impact of using the AI campaign module was significant, increasing the number of deposits by 53.75% and the amount committed to term deposits by 50.48% (treatment group compared to the control group). The fall campaign generated $53.9 million in total over four weeks, having benefited from self-learning mechanics. The AI campaign module was responsible for increasing the number of deposits by 46% and the amount committed to term deposits by 45% (treatment group compared to the control group).

We applied a similar methodology to identify a group of customers who had a higher chance of renewing their mortgages, thereby prioritizing the efforts of loan officers on high-risk customers. By using propensity models and incorporating behavioral features derived from five years of financial and credit records, we aimed to enhance the efficiency and effectiveness of the loan officers' outreach.

The model prioritized customers into several groups based on their likelihood to renew their mortgages. This segmentation enabled loan officers to focus their efforts on those with the highest propensity to renew, ensuring a more targeted and efficient use of their time and resources. After implementing and monitoring the actual performance of the call center staff, the results were compelling. The prioritized lists generated by the propensity models led to a 40% efficiency gain.

These applications of propensity models illustrate their powerful impact on targeted marketing and customer engagement strategies. By accurately identifying high-risk and high-potential customers, organizations can prioritize their efforts, optimize resource allocation, and achieve significant efficiency gains. The strategic use of propensity models, combined with deep behavioral insights, enables businesses to move beyond generic outreach efforts and engage customers in a more personalized and effective manner.

This approach exemplifies how advanced analytics can drive substantial business outcomes, enhance operational efficiency, and foster stronger customer engagement. As we continue to refine these models and integrate more data sources, the potential for even greater impact and innovation in customer engagement strategies becomes increasingly apparent.

Personalized Pricing: Influencing Behaviors and Financial Outcomes

Another area where the combination of behavioral economics and ML algorithms can effectively influence behaviors and financial outcomes is in reimagining the delinquent customer journey to reduce defaults.

During the pandemic, we partnered with one of the largest providers of credit cards in North America, primarily serving the subprime population. This key player in the finance industry, with 3 000 employees and an annual revenue of over $1 billion, faced significant challenges.

The senior executives were keen on creating substantial value by improving debt collection strategies and reimagining the delinquent customer journey through advanced analytics, behavioral sciences, and ML. At that time, they were experiencing a very low collection rate of about 8% by sending generic settlement offer letters to everyone in delinquency and charge-off.

This low collection rate was not too surprising for us: The risk attitudes of individuals with low incomes and low credit ratings differ significantly from those in higher socioeconomic tiers. These individuals might accept offers to settle their debt over, say, 24 months and initially make monthly payments. However, if they encounter a financial "shock" – such as a child falling ill and incurring high medical bills or a spouse losing their job – they will prioritize immediate necessities over debt payments. Consequently, the "survival rate" of these payment and settlement offers is often too low. One-size-fits-all solutions fail to account for the varying risk profiles of these customers and the factors driving their behaviors.

Our first step was to determine if we could predict which customers were at risk of becoming delinquent or defaulting. Delinquency occurs when a borrower fails to make a payment on time, whereas default is a more severe state where the borrower has missed multiple payments and is considered unlikely to meet their debt obligations. We aimed to identify the most critical predictive signals behind these risks.

We began by analyzing a combination of factors such as "missing payments," "not sufficient funds (NSF)," "credit ratings," and "frequency and timing of spendings" to understand their correlation with delinquency and defaults. Additionally, we examined how these factors related to socioeconomic demographics. Our goal was to identify if there was a significant difference between those who fell into delinquency but eventually resumed making payments and those who obtained the credit card with the intention to default from the start. We sought to differentiate these behavioral signatures.

For instance, a customer with a pattern of regular spending followed by sporadic NSF might exhibit a different risk profile than someone whose credit rating steadily declines. We also looked at how external factors, such as local economic conditions or unemployment rates, affected these behaviors. Understanding these nuances was critical in tailoring our approach.

By now, it shouldn't be surprising to you that we also conducted a thorough behavioral segmentation to uncover hidden traits that helped group customers. We used XGBoost and propensity models to build several predictive models that accurately identified which customers were at risk of delinquency and which delinquent customers were at risk of defaulting. Implementing a propensity score enabled us to proactively identify these high-risk customers, helping to control financial risk and design interventions aimed at increasing their financial well-being.

We went a step further: For individuals already in default, we designed personalized settlement offers to maximize the probability of offer acceptance, survival, and payment plan completion. Consider this example: An individual who has missed two to three consecutive payments receives a settlement offer with the following options. Let's say the amount owed is $1200.

1. **Settle immediately and receive a 50% discount:** The borrower can settle today by paying only 50% of the debt, which in this case would be $600 on a $1200 debt.

2. **Settle within a year by making 12 payments and receive a 35% discount:** The borrower can pay 65% of the debt, totaling $780, over 12 payments, equating to $65 per month.

3. **Settle within two years by making 24 payments and receive a 20% discount:** The borrower can pay 80% of the debt, which is $960, over 24 months, resulting in monthly payments of $40.

Which option would you choose?

Let's analyze these options in more detail. At first glance, option one seems the most attractive because it offers the greatest discount, allowing the borrower to pay $600 and settle the debt immediately without significantly affecting their credit rating. However, not all customers have the liquidity to make such a lump-sum payment, making the second and third options more viable for some.

The second option provides a manageable monthly payment of $65 over a year, and the third option extends the repayment period to two years with even lower monthly payments of $40. Each option is designed to cater to different financial situations and preferences, acknowledging that customers facing financial difficulties need flexible solutions.

Behavioral Economics in Personalized Pricing Strategies

Of course, if we were Econs, we would choose option 1 because it is the logical choice to save the most. But we are not! We have discussed loss aversion in previous chapters: we hate losing more than we love gaining the same amount. On top of that, we also discount the future. This means that future gains are perceived as less valuable than immediate gains, even if they are the same amount, and future losses are also perceived as less painful.

Real-World Impact

Additionally, consider the socioeconomic and risk profile of this group of individuals – they are in the subprime category. Even if they are very logical, they can't settle at a 50% discount if they don't have $600 on hand now, or they'd be short for the rent at the end of the month.

So, you might be thinking, all right, what is the problem with the third option? It is better for the bank anyway – they'd collect 80% of the owing amount, and the individual gets to make smaller regular payments every month. True, except that the survival rate of these longer-term payment plans is very low. This means that individuals only make regular payments for the first few months and then become more irregular in making payments and eventually stop altogether. Why does this happen? Let me ask you a different question: How often do you start an exercise routine or a diet plan or simply take vitamins with the highest conviction and best intentions, only to abandon it after a few weeks or months? We all know exercise, diet, or taking vitamins is very good for us. My mom reminds me of this daily! It is not about being unaware of the benefits. But why do we have a hard time sticking to it? It is all about loss aversion all over again: you must do a costly activity now and only realize the gains in the future. But wait, we also discount the future, so the gains feel less significant because they are in the future!

The same is true here, and why the survival rates are much lower in a 24-month settlement term than in a 12-month one. There are also no real perceived gains for the individual other than not damaging their credit score. This is obviously a very important factor that influences future access to credit, mortgages, and other types of loans. But an individual in the subprime category already has a low credit rating, so the "reward" of completing their payment plan might not be as motivating.

To address this, we need to design settlement offers that align better with the behavioral tendencies of the customers. Shorter-term plans with immediate, tangible benefits, such as partial debt forgiveness after a few months of consistent payments, can increase the likelihood of adherence. Additionally, offering small incentives for early payments or bonuses for maintaining regular payments over shorter periods can make the repayment process more engaging and rewarding.

The question for us was whether we could estimate and predict the survival rate of these offers for different individuals based on their past payments and other behavioral traits. That is exactly what we did. We built a set of models to predict the probability of each individual accepting a certain offer. We also estimated the probability of survival for each payment plan. The idea was to find a settlement plan that would maximize both the acceptance rate and the survival rate, thereby increasing conversion rates and debt recovery. To minimize choice fatigue and the effects of loss aversion, we simplified the settlement offer letters to present only two options. These optimized and personalized settlement plans led to a 65% higher expected recovered debt.

Personalized pricing and offers also work effectively in the context of retail. Different individuals experience varying levels of price sensitivity at different times. For instance, retail sales are typically higher at the beginning of the month than at the end, influenced by factors like paycheck cycles. A 15% discount might be enough to motivate my aunt, but less than a 50% discount won't be enough for me to voluntarily go to a mall. That said, my friend and I pass 25% discount coupons to each other for Ralph Lauren because they don't go on discount that often. The same is true for Jimmy Choo shoes! Another example is when a $200 discount would get me to go shoe shopping, but a $200 discount won't incentivize me to buy a new laptop or phone. How our biases and heuristics motivate us to make funny and irrational decisions is fascinating, and even more fascinating is how understanding what makes us tick can be used to guide our choices and decisions.

For further insights into how behavioral economics can be applied to influence decisions, I highly recommend reading *Nudge* by Richard Thaler. This book delves deep into how subtle nudges can shape behavior, making it a compelling read for anyone interested in the intersection of economics and human psychology. By harnessing these insights, businesses can create strategies that not only drive financial outcomes but also align with the natural tendencies of their customers, fostering loyalty and long-term engagement.

Forecasting: Understanding the Dynamics of Demand

Have you ever encountered a scenario in which product prices rise, and instead of demand dropping, it actually increases? It's counterintuitive, isn't it? No one typically thinks, "I'll wait for the price to go up before I buy." Yet, many retailers and consumer goods companies witnessed this very phenomenon during the pandemic.

Consider the concept of positive price elasticity we discussed previously. Let me refresh your memory with a case from a client of ours during the early months of COVID-19: Due to supply chain disruptions, they had to raise the prices of their bread and bakery products. To their astonishment, sales surged. When they repeated the price hike a few months later, sales increased again. Buoyed by this apparent success, their executives decided to implement another price increase at the start of the pandemic's second year, believing they could capitalize on this positive price elasticity. As market leaders, they assumed that higher prices signaled better quality and helped them outperform the competition.

However, they overlooked a crucial detail: During the early pandemic months, restaurants were closed, and people worked from home, leading to increased home consumption. Consumers also had more disposable income and didn't mind moderate price increases. These factors combined to boost demand.

The third price hike decision was fraught with greater risk. If the demand surge was genuine and enduring, they needed to ramp up production, and potentially build a new facility. But if the spike was temporary and demand flattened as the world adapted to a new normal, they would need to adjust prices back down or possibly lay off the extra labor they had hired.

This highlights the importance of accurate forecasting in understanding and anticipating market dynamics. Forecasting models that consider a broader range of variables – economic conditions, consumer behavior shifts, and external factors like a global pandemic – can provide more reliable insights. Predictive analytics and ML can help decipher these complex patterns, enabling companies to make informed decisions about pricing, production, and resource allocation.

Accurately estimating demand is one of the most critical tasks faced by consumer goods companies, retailers, and brands. Ineffective demand forecasting can lead to either excess inventory or underproduction, eroding margins and leaving significant revenue on the table. This task is further complicated by the lack of proper data, often due to siloed or outdated technologies. Most consumer packaged goods (CPG) companies do not have direct access to consumer data and must rely on purchasing POS data or use their shipment data as a proxy for demand data. This is far from ideal, because it lacks information on the inventory levels of their customers.

To build effective demand forecasting models, a few years of transactional or shipment data containing quantities sold or shipped and prices is necessary. When a major shock like a pandemic occurs, it becomes crucial to control for observable factors that might affect consumer behaviors, such as interest rates, seasonality, and competitive product prices.

This is precisely what we did for our client. We explored all available data at the most granular levels, controlling for various macro and micro factors, and demonstrated that the price elasticity of consumers was indeed negative. This enabled us to construct a robust demand forecasting model based on XGBoost, as opposed to simple regression or time series models. The results were impressive, achieving over 94% accuracy with aggregated weighted errors of less than 3%. Previously, their forecasting accuracy hovered around 75%. By leveraging advanced predictive analytics, our client was able to navigate the complexities of consumer behavior, even amid unprecedented disruptions, ensuring a more resilient and responsive supply chain.

To demonstrate why forecasting tasks are so difficult, let me contrast this with another scenario in which we were not as successful. During the same time frame, we worked with another client who aimed to improve their shipment forecasting from their distribution centers to their customers, primarily grocery stores. In other words, they needed to know how much each customer would order at least four weeks in advance to ensure they had the right amount of inventory and supplies at their distribution centers. They only had access to POS and demand data for half of their distribution centers from other data providers, and unfortunately, it didn't perfectly match their internal shipment and price datasets.

Data on discounts and promotions is another crucial factor in building successful forecasting models because discounts drive higher sales volumes. Although they had this data available, the granularity level was challenging to match with the rest of the data, especially because promotions were sometimes combined with other discounts and marketing activities. Adding to the complexity, there were a series of promotions introduced during the pandemic that didn't exist in the historical data.

We tried many different methods and models to build the most accurate shipment forecasting model, and our models were as good as what their sales teams and account managers were forecasting, but we couldn't increase the accuracy beyond a point. The moral of the story is simple: Consumer behavior and preference change rapidly and dynamically due to macro conditions among other factors. Without collecting granular and high-quality data, even some of the best models won't be able to produce high-accuracy outcomes.

The most fascinating thing, although not too surprising, was when we took the same model architecture to a different, more complex environment with an order of magnitude more data, we predicted the demand with more than 20% higher accuracy than their baseline of 65%. The first set of outcomes from our model, trained only with the MVD, produced an average of 80–85% accuracy across five different product categories. What was perhaps more astonishing was that we did that in less than a month.

This comparison highlights the critical importance of high-quality, granular data in achieving accurate forecasting. In environments where data is rich and well-integrated, advanced models can significantly enhance predictive accuracy and provide valuable insights that drive strategic decision-making. Conversely, in scenarios where data is fragmented or insufficient, even the most sophisticated models struggle to deliver precise forecasts. The key takeaway is that the success of forecasting efforts hinges not only on the sophistication of the models but also on the quality and comprehensiveness of the data they are built on. As the saying goes in ML, "garbage in, garbage out." Understanding and addressing these data challenges is essential for leveraging the full potential of predictive analytics in any industry.

The Power of Forecasting and Optimization

One reason why achieving high accuracy in forecasting is important is because of the level of optimization it enables as a result. Sometimes I hear amusing statements from executives, such as "We have a very good demand forecasting engine, but we still have a lot of excess inventory!" Or they wonder how they can optimize marketing spend while knowing they don't have a good handle on the return on investment (ROI) of their marketing activities. I will come back to demand forecasting and inventory optimization in a moment. But let's talk about marketing optimization first.

When we were trying to solve the positive price elasticity puzzle, I was introduced to another strange phenomenon that happens in retail, especially in the CPG sector. Because most CPGs don't have first-party consumer data, they usually outsource their marketing spend to an agency. The agency then uses mixed marketing models (MMMs) – statistical analysis techniques that estimate the impact of various marketing channels on sales – to calculate and estimate marketing ROI per channel. What makes it strange is not just the lack of clarity in the assumptions, such as marketing saturation, but the fact that marketers send this data out at the end of the year and receive the results six months later. By then, they have already decided on their marketing spend and strategy, making the results either a confirmation bias or simply ignored.

This disconnect between data analysis and actionable insights is not just a minor inconvenience; it's a significant barrier to effective optimization. Imagine a scenario in which you could receive real-time insights on the impact of your marketing spend. You could adjust your strategy dynamically, reallocating resources to the channels delivering the highest ROI and pulling back on those that aren't performing. This is the promise of advanced analytics and real-time data processing, which can transform how marketing decisions are made.

Another strange, but more understandable, fact is that marketing input and spending never enters the demand forecasting models. Marketing, by definition, is supposed to drive behavior and demand, and yet it is the only factor not used in forecasting demand. The reason is that marketing activities are usually highly correlated with promotions. If marketing spend naively enters the demand forecasting models, there will be multicollinearity – a situation in which predictor variables are highly correlated with each other. This increases the noise in the system and lowers the accuracy. We learned this the hard way.

Transparent MMMs

The first challenge was to build an MMM with transparent assumptions to help our client understand the effectiveness of their marketing activities in real time. We spent considerable time understanding the context of their business, the cadence of their marketing activities, the overlap with other promotional and merchandising efforts, and the anecdotal evidence about what works and what doesn't. We also consulted with marketing professors and conducted extensive literature reviews.

Ultimately, we developed an MMM based on Bayesian optimization that adjusted assumptions on saturation curves and ad-stocking, which establishes how long the effect of advertising lasts depending on the marketing channels and other factors. This model enabled the client to perform experiments and compare the ROI of different marketing channels and activities without waiting several months, allowing for timely strategic adjustments. The marginal ROI curves also helped the client discern which channels might be more impactful.

We then decided to push the boundaries further. We used the model to simulate various scenarios and estimate the ROI for each. For example, they could input their entire marketing spend plan for the next year into the simulator, and the model would predict the ROI per channel.

Ensembling Models for Enhanced Forecasting

Next, we were curious if by ensembling the MMM with demand forecasting models, we could improve the accuracy of the forecast. Ensembling is an ML technique that combines multiple models to produce a more accurate prediction. This approach proved successful: marketing spend explained 42% of the unexplained errors in our forecast, increasing the accuracy of the system to 96.5%.

This improvement was not just about enhancing accuracy; it enabled us to properly consider the impact of marketing when optimizing prices or deciding how much to spend on merchandising versus advertising on TV or social media.

In any business, resources are always limited, and there are trade-offs in budget allocation across different activities. It is indeed possible to leverage the power of data to help de-risk decisions. Combining information from all parts of the business into one streamlined system is essential for proper optimization. This integrated approach allows for more informed decision-making, ultimately driving better business outcomes.

The Interplay of Demand Forecasting and Inventory Optimization

Demand forecasting and inventory optimization go hand in hand. It's impossible to have a bad demand forecast and still achieve efficient levels of inventory, or to have a good demand forecasting system that leads to excess inventory – assuming store managers or vendor relationship managers aren't overwriting the ML system's outputs.

Imagine a retailer with over 80 grocery stores and pharmacies carrying more than 100 000 products and SKUs. Forecasting demand for these products, considering competition and cannibalization effects, is a very difficult task. Some products might sell over 100 units in a day, and others might sell only 10 units over a month or a year. Add seasonal items and discontinued or substituted products to the mix, and the complexity increases exponentially. Many retailers take a hybrid approach, forecasting some items manually and others using forecasting models. The logic behind some of these models is also very simplistic, often just picking an average number based on monthly sales or using a min-max approach.

When we started working with this grocer, fewer than 20 000 items were being forecasted using an NN model with only one layer.[5] This simplistic approach, designed by a large consulting firm, took over 12 hours to run the data pipelines and produce predictions, making it computationally very costly. The data pipelines were inefficiently designed. We optimized them, reducing the runtime to less than a minute, and built a more sophisticated model with seven NN layers, enabling high-accuracy forecasts for over 60 000 items.

Once we had improved demand forecasting, we turned our attention to the inventory optimization problem. Unfortunately, partly due to poor demand forecasting, the retailer had accumulated excess inventories, with storage costs eroding margins. Inventory optimization models are statistical models that help identify optimal minimum and maximum levels of inventory based on demand forecasts. When the stock in the store or distribution centers hits the minimum levels, the system is triggered to order more supplies.

The challenge with most of these statistical models is twofold: first, the demand forecasts are often inaccurate, and second, the assumptions used in these models are overly simplistic. For example, they might assume a fixed or uniform distribution of demand. After improving the demand forecasting system and refining the inventory optimization models, our simulations showed that the new system could save up to $13 million in a quarter compared to the previous models.

Conclusion

In this chapter, we've explored various applications of ML and AI, emphasizing the importance of integrating behavioral economics to understand deeper connections and drive meaningful outcomes. From advanced recommendation engines and targeted campaigns to personalized pricing and demand forecasting, the fusion of AI with behavioral insights has demonstrated significant potential.

Building effective ML and AI models requires not only robust data but also a comprehensive understanding of the behavioral factors influencing consumer decisions. The examples we've discussed illustrate how nuanced behavioral signals, when combined with sophisticated algorithms, can lead to optimized decision-making processes, enhanced customer experiences, and substantial financial gains.

As we move forward, it's essential to recognize that the power of AI and ML lies not just in their technical capabilities but also in their ability to integrate seamlessly with human insights and organizational goals. The journey doesn't end here; it's just the beginning of unlocking the full potential of AI and ML in the dynamic landscape of business.

CHAPTER

6

Decoding Complexity
Leveraging Systems Thinking in Modern Organizations

"The measure of intelligence is the ability to change. . . . Anyone who has never made a mistake has never tried anything new."

– Albert Einstein

In 2022, I found myself seated next to a senior executive from the world's largest pharmaceutical and supply chain company on an evening flight from Las Vegas to New York City. Fresh from speaking at a major retail conference, I was engrossed in my work when his curiosity was piqued – perhaps by the fervor of my typing. After a brief introduction, he peppered me with insightful and direct questions about my background and the work my company does. Despite my initial reluctance to engage in a late-night conversation after three intense days at the conference, his questions were intelligent and intentional, revealing a genuine interest.

I shared my bewilderment over enterprises spending hundreds of millions – sometimes billions – of dollars on "organizing and cleaning data" or the frequent failures of enterprise resource planning systems during implementations. I highlighted how, even after years of investment in data and foundational capabilities, organizations often barely scratch the surface of using data and analytics, let alone extracting insights from predictive models. Fundamental questions about customers were overlooked, demand forecasting was mishandled, and supply chain disruptions led to customers missing products when they needed them most.

Unbeknownst to me, he was the global head of supply chain for this multinational organization, and my candid thoughts were music to his ears. A few months later, we began working with his organization, focusing on predicting customer experience disruptions due to supply chain issues. The aim was to proactively engage and communicate with customers most at risk of experiencing delays exacerbated by the pandemic.

We decided to start our project in the United States, leveraging available data on past orders, complaints, disputes, calls, and a comprehensive history of disruptions such as back orders, split deliveries, and long delivery times. This dataset, encompassing hundreds of thousands of customers and millions of orders over the past five years, formed the backbone of our recommendation model. This model functioned as a demand sensing tool, predicting what each customer would order one month and one quarter in advance at a granular level.

Next, we developed another set of models to analyze the predicted items in the customers' baskets for the upcoming month and quarter. These models assessed whether each predicted item would face any form of supply chain disruption. The rationale behind this prediction time frame was straightforward: If we could accurately forecast what customers would order and whether these items would be delivered on time and in full (OTIF), we could identify the key signals and reasons behind potential disruptions. This foresight would provide the delivery team with ample time to make necessary adjustments or collaborate with the commercial and account teams to proactively communicate with customers, mitigating any negative impact on their experience.

To pilot our initiative, we decided to focus on the five largest and most strategic accounts in the United States. This turned out to be our first mistake. For each account, we conducted strategy sessions with account managers, commercial leaders, delivery leaders, supply chain experts, and analytics leaders to review the results and understand the specific business constraints. We also decided to take action only on items and orders with 95% accuracy or higher. This approach seemed logical: the account executives were risk averse and hesitant to communicate potential issues to their top customers if there was only a 75% probability of occurrence.

Initially, everything appeared to be on track. We systematically addressed the various concerns and doubts raised by the account executives and commercial leaders. We conducted thorough behavioral segmentation and demonstrated how different metrics correlated with customer satisfaction, complaints, disputes, and even order sizes among other financial metrics. The next logical step was to run a pilot based on the predictions for the next quarter.

However, this did not come to fruition. Despite the compelling evidence we presented, the team was uncomfortable reaching out to their customers to give them a heads-up about potential challenges. The reluctance to act, driven by the fear of upsetting their best customers or being unable to provide concrete solutions, stalled the implementation of our proactive strategy.

Their reluctance stemmed from fears such as, "What if they get upset?" or "What if they ask us what we are going to do about it?" Moreover, there was no data to quantify the cost of poor customer experiences due to consistent disruptions. Although multiple customer surveys indicated that proactive communication could improve experiences, the teams still hesitated.

Here we were, having built several models with high accuracy at the most granular level, enabling the organization to proactively communicate with their customers. Yet, they simply refused to act. What was happening? Didn't they care about their customers? Of course, they did. The hundreds of hours we spent together were a testament to their dedication to better serving their customers.

I should have anticipated this hesitation; after all, I was the behavioral economist in the room! Let us analyze these events step by step.

Only a Wet Baby Likes Change: Loss Aversion + Status Quo Bias

Our first mistake was choosing the most strategic and largest accounts, managed by the most senior and experienced account executives. These accounts were perpetually under the spotlight. Senior executives had initially suggested these accounts because they aimed to enhance the experience of their most important customers, thereby ensuring their loyalty and increasing engagement, which would, in turn, drive revenue and margin growth. After all, improving OTIF deliveries, as well as reducing complaints and disputes, would positively affect payment terms, improve cash flow, and increase margins.

However, the high visibility and importance of these accounts made the executives risk averse. They were hesitant to implement changes or communicate potential issues with their most valuable customers, fearing any negative repercussions. This reluctance ultimately hindered our efforts to proactively address supply chain disruptions and improve customer satisfaction. The very factors that made these accounts critical also made them resistant to the innovative approaches we proposed.

The problem lay in the fact that if anything went wrong, the senior executives were not held accountable; it was the account executives and their commercial partners who bore the brunt of the responsibility. The issue wasn't a misalignment of incentives but rather an asymmetry in risk exposure. This imbalance exacerbated the effects of cognitive biases like loss aversion and status quo bias.

Status quo bias, a well-documented cognitive bias, stems from our preference to maintain the current state of affairs, using it as a baseline against which we measure any deviations. It acts as our default option. Extensive research has shown how status quo bias frequently affects decision-making, intertwining with other biases such as loss aversion, the endowment effect, longevity, regret avoidance, and psychological commitment. These biases collectively create a powerful force that resists change, even when that change could lead to significant improvements.[1]

From loss aversion, we know that the psychological pain of losing is more powerful than the pleasure of gaining. When combined with status quo bias, individuals tend to weigh the potential losses of deviating from the status quo more heavily than the potential gains. This leads to greater regret for actions that change the status quo than for those that maintain it. Consequently, these forces favor the status quo, motivating people to do nothing or to stick with current or past decisions. Change is avoided, and decision-makers tend to adhere to what has been done before. In essence, we resist change unless the perceived benefits significantly outweigh the risks.[2]

Our situation was even more challenging because, despite our best efforts, it was impossible to quantify the cost of supply chain disruptions and poor customer experiences. All the evidence we had was anecdotal, and there were no datasets that could directly correlate different aspects of the supply chain with lost revenues or additional costs incurred by our client. For instance, although it is widely acknowledged that express delivery packages are significantly more expensive when deadlines are missed, no one could specify by how much. This lack of concrete data made it difficult to quantify the benefits of proactive communication against the perceived risk of upsetting the customer with potential bad news, even though proactive communication was precisely what customers desired.

This inability to present a clear, data-backed cost–benefit analysis exacerbated the status quo bias and loss aversion among the decision-makers. Without tangible figures to illustrate the financial impact of supply chain failures or the advantages of timely communication, the account executives and their teams were more inclined to stick with their existing practices. The fear of making a wrong move, compounded by the asymmetrical risk environment, made them resistant to change. Even though proactive communication could have improved customer satisfaction and loyalty, the psychological weight of potential losses overshadowed the potential gains, leading to inaction and a continuation of the status quo.

Information limitations and asymmetries often underpin the logic behind status quo bias. Decision outcomes are rarely certain, and their potential utility is even less so. Some errors carry higher costs than others,[3] making it seem safer to stick with past decisions as long as they have been "good enough." The cognitive load and mental effort required to evaluate the outcomes of different choices in an uncertain environment often make the status quo more appealing, simply because its outcomes are easier to predict. No one was ever fired for not proactively communicating with customers; that was just the way things were done.

Moreover, the longer a status quo has existed, the harder it becomes to change it.[4] This is due to longevity and existence biases, which lead individuals to favor options that have been around for a long time. The more entrenched something is, the more people perceive it as better or more valid, making change increasingly difficult. This phenomenon mirrors the evolutionary concept of "survival of the fittest,"[5] where long-standing practices are seen as inherently superior.

In our case, the absence of concrete data to quantify the costs of supply chain disruptions and the benefits of proactive communication exacerbated this bias. Without solid figures to illustrate the financial impact or the advantages of timely intervention, the decision-makers were naturally inclined to maintain their existing practices. The fear of potential losses, combined with the cognitive effort required to assess new strategies, made them reluctant to deviate from the established norm. Consequently, the perceived safety and familiarity of the status quo continued to overshadow the potential benefits of change.

To summarize, the risk and perceived cost of making a mistake were too high for the account executives we were working with to deviate from the status quo. This reluctance was compounded by our inability to quantify the benefits of proactive communication or to suggest minor adjustments, such as changing order dates from Thursdays to Tuesdays to avoid weekend delays. Additionally, this was the entire team's first experience with predictive models (even though demand plans technically came from a demand forecasting model managed by another group). Although they understood the concept of probabilities and predictive models, they were expecting 100% certainty. This insistence on absolute certainty further entrenched their resistance to change.

Loss aversion and status quo bias can be incredibly powerful tools hindering innovation. I am reminded of what one of my dear mentors and advisors used to tell me: Only a wet baby likes change.

Why did I not see all this, as obvious as it seems in hindsight? Because I also have my own biases. The one most at work was the "curse of knowledge" or "curse of expertise." This cognitive bias occurs when someone, assuming others share their knowledge and understanding, fails to recognize that others lack the same information. The term "curse of knowledge" was coined in a 1989 *Journal of Political Economy*[6] article by economists Colin Camerer, George Loewenstein, and Martin Weber. They demonstrated that more-informed individuals often cannot accurately predict the judgments and behaviors of less-informed individuals.

I only began to grasp the extent to which loss aversion and status quo bias were influencing the team when it was almost too late. The strong desire to avoid potential losses and the comfort of sticking with established practices were powerful deterrents to change. Recognizing this, we decided to pause the pilot, rethink our approach, and start with a different part of the organization.

Reflecting on this experience, I realized the importance of addressing cognitive biases head-on. Understanding that the team needed not just data, but also reassurance and a gradual introduction to predictive models, was crucial. By acknowledging and mitigating these biases, we could better foster an environment open to innovation and improvement. This lesson underscored the need for patience and empathy when driving organizational change, ensuring that all stakeholders are on board and comfortable with the transition process.

It Gets Better! Commitment Device, Peer Effect, and Sunk Cost Fallacy

We were not going to give up that easily. In the first few months of the project, we engaged with executives in similar roles but in a completely different part of the business. We decided to run a small pilot project alongside our main initiative in a different market: Canada. Instead of predicting demand, we focused directly on predicting supply chain disruptions and quantifying their impact on customer experience.

We conducted a thorough behavioral segmentation, identifying at-risk customers and analytically demonstrating the reasons behind their risk status. This approach enabled us to showcase clear, data-driven insights into potential disruptions and their effects. At this point, we had not yet faced the challenges of operationalizing in the US market with the top 10 strategic customers.

By targeting a different segment and concentrating on a specific aspect of the supply chain, we aimed to build trust and demonstrate the value of predictive models. This strategy not only provided us with valuable insights but also helped us refine our approach. The Canadian pilot project served as a proving ground, enabling us to test our assumptions and methodologies in a less pressured environment.

We were halfway through the pilot in the United States when another business unit decided to scale the work we had done in Canada and expand it to other markets, particularly in Latin America. By then, I had learned some hard lessons, and we set out to approach things differently this time.

Our first step was to identify all relevant groups and departments. We conducted interviews with individuals from five different departments, ranging from analysts to the presidents of the business units. This comprehensive approach enabled us to map out dependencies and understand the flow of information. We also gained a much clearer picture of how decisions were made.

We recognized the importance of understanding the risk environment and saw how involving a number of people in the decision-making process could lower perceived risk. By having various individuals commit to a particular path, accountability was shared, reducing the likelihood of finger-pointing if things went wrong. This approach also built trust quickly because it demonstrated that we were consulting people at all levels and taking their input seriously.

Our goal was to reduce information asymmetries as much as possible, ensuring that our key stakeholders felt heard and that their pain points were addressed. By fostering an inclusive and transparent process, we aimed to create a more collaborative and supportive environment for implementing and scaling our predictive models.

Beyond gaining their trust, we needed to secure their commitment to a pilot right from the start, knowing that loss aversion and status quo bias would inevitably set in. However, it's challenging to commit to something uncertain. There was no guarantee that the models would perform well or generate actionable insights. This was also their first experience with predictive models at this scale. So, we did something that data scientists usually dislike: We established multiple weekly touchpoints with different groups to review the approach, results, findings, challenges, and more on a very regular basis. We asked them to commit to these meetings.

Maintaining this frequency of meetings was difficult for my team, especially while combing through gigabytes of data from

50 different datasets and building 18 different machine learning algorithms to capture various dimensions of supply chain disruptions and customer experience. This was hard work that required meticulous attention to detail, and no data scientist enjoys showcasing half-done models that do not yet perform well. Additionally, we essentially put the client's team through a statistics and machine learning course during the four months of building and training these models.

This frequent interaction and education not only kept the client engaged but also helped them understand the complexities and potential of the predictive models. By fostering a collaborative learning environment, we aimed to build their confidence in the process and ensure their ongoing commitment. This approach, although demanding, was crucial for overcoming initial resistance and paving the way for successful implementation.

Let me highlight a few key points: First, the internal teams gradually began to appreciate the complexity of these models and methods. They observed our meticulous process of testing every hypothesis, cleaning data, and building the models. More important, they saw how their suggestions, thoughts, and curiosities were tested or incorporated into our work. We also pointed out some of their biases from time to time. For instance, they held certain beliefs about Brazil that we found no data-driven evidence for, or some insights from the model were not intuitive to them. In such cases, we either demonstrated why the data supported those insights or, occasionally, sought additional data to conduct further tests.

This collaborative approach fostered a deeper understanding and trust in the predictive models. By actively involving them in the process and addressing their biases, we built a strong foundation for effective decision-making and ensured that the models were not only technically sound but also aligned with their practical needs and expectations.

This continued for several months. Despite the difficulty and unfamiliarity of the territory, they remained committed. Why? Three factors worked in our favor this time: commitment device, peer effects, and the sunk cost fallacy. Let me explain: Instead of asking them to commit to a pilot, we simply asked them to commit to a series of meetings. This seemed manageable to them; it was time-consuming but not nearly as risky as committing to running a pilot. They also had their colleagues and friends at the meetings, where they were learning or discussing something interesting and different. No one wanted to be the individual or group left out, showcasing the power of peer effects[7] and the fear of missing out. Thus, the commitment device[8] worked due to its low risk, engaging environment, and the influence of peer dynamics.

Moreover, as they continued to invest time and effort into these meetings, the sunk cost fallacy played a role. Having already committed significant resources, they were motivated to see the project through to fruition, not wanting their initial investment to go to waste. This combination of low initial commitment, engaging peer interactions, and the psychological pull of sunk costs kept them engaged and invested in the project, ultimately fostering a collaborative and productive atmosphere.

Let us talk about sunk cost fallacy a bit more: The more time they spent with us, the more they wanted to see the project through. They shared more ideas, requested more tests, and became increasingly committed. It reached a point where we had to ask them to slow down because we could not boil the ocean in four months! This was the sunk cost fallacy at work – in a positive way. The sunk cost fallacy occurs when individuals are reluctant to abandon a strategy or course of action because they have already invested heavily in it, even when it's clear that changing course would be more beneficial.

Have you ever been to a movie where it was clear within the first 30 minutes that it was not good? Why do we sit through the whole film? We tell ourselves, "Let us watch 15 more minutes," but by then we have been at the theater for 45 minutes, and we think, "What if it gets better in the next 15 minutes?" And so on. The more we invest time and resources, the more committed we become, making it harder to change course.

We leveraged this dynamic to our advantage, encouraging continued engagement and commitment. However, it wasn't just the sunk cost fallacy at play – the results were very promising. The combination of their growing investment and the positive outcomes created a reinforcing cycle of commitment and enthusiasm, ultimately driving the project forward successfully.

This does not mean we did not encounter challenges as we approached the pilot phase. The team frequently reconsidered the metrics for measuring customer experience; some metrics agreed on the previous month were deemed insufficient and needed revision. At one point, someone even expressed a desire for 100% accuracy, prompting us to gently remind them of our extensive discussions on predictive versus deterministic models.

However, their queries took on a different tone. Although still concerned with the risks of running a pilot, they were now focused on ensuring that all relevant factors had been considered, rather than trying to prove that we had overlooked something. They had invested as much time and effort as we had and were equally committed to seeing the project through to success. This shift in perspective highlighted their growing confidence and engagement with the project, reinforcing the collaborative spirit that was crucial for its eventual success.

So, we worked even more closely with them for another three months, this time focusing on enabling and helping them to conduct the validation and tests internally. This approach enabled them to convince themselves that all the *i*s were dotted and all the *t*s were crossed. This was fine with us, although my data scientists were a bit frustrated – remember the "curse of knowledge" or the "curse of expertise"?

The more time they spent on the project, the more committed they were to seeing it succeed. The peer effect was also at play because multiple local and global teams were involved and overseeing the validation process. They even assigned an individual to work internally on building dashboards and making it easier for the teams to conduct validations. This was an excellent development!

After three months of internal validation, we were ready to move forward, make necessary changes, and help the team activate and monitor the results.

The success of this project was driven by several key factors. First, we recognized the importance of addressing cognitive biases such as loss aversion and status quo bias, which initially hindered progress. By involving multiple stakeholders from the outset and fostering an inclusive environment, we reduced information asymmetries and built trust. Frequent touchpoints and collaborative meetings kept everyone engaged and aligned, while the peer effect and fear of missing out maintained commitment.

Moreover, our willingness to educate the client team and involve them deeply in the validation process turned potential resistance into active support. By allowing them to take ownership of the testing and validation phases, we ensured that all parties were confident in the project's outcomes. This iterative approach, combined with transparency and continuous communication, ultimately paved the way for a successful implementation.

These strategies not only addressed the inherent challenges but also created a robust framework for ongoing collaboration and improvement. The project's success underscores the importance of aligning technical expertise with organizational dynamics, fostering a culture of trust, and maintaining a flexible, adaptive approach to innovation.

Conclusion

The journey from a chance encounter on a flight to implementing a predictive model for supply chain disruptions was both challenging and enlightening. Our initial attempt to address supply chain issues with the organization's most strategic accounts revealed deep-seated cognitive biases, such as loss aversion and status quo bias, which significantly hindered change despite the potential benefits. The high visibility and importance of these accounts made the executives risk averse, preventing proactive measures that could have mitigated supply chain disruptions and improved customer satisfaction.

Reflecting on this experience, it became clear that understanding and addressing cognitive biases are crucial in driving organizational change. Loss aversion and status quo bias can create formidable barriers, even when innovative solutions promise substantial improvements. Our inability to present a clear, data-backed cost–benefit analysis further exacerbated these biases, highlighting the need for concrete evidence to support change initiatives.

However, by shifting our focus to a smaller, less pressured market, we were able to build trust and demonstrate the value of our predictive models. This experience highlighted the critical role of adaptability and understanding behavioral economics in driving organizational change and maintaining engagement and driving projects forward.

Our iterative approach, characterized by frequent touchpoints, education, and collaborative validation processes, transformed initial resistance into active support. We established multiple weekly touchpoints with different groups to review the approach, results, findings, and challenges. This frequent interaction and education not only kept the client engaged but also helped them understand the complexities and potential of the predictive models.

Moreover, our willingness to involve the client team deeply in the validation process turned potential resistance into active support. By allowing them to take ownership of the testing and validation phases, we ensured that all parties were confident in the project's outcomes. This iterative approach, combined with transparency and continuous communication, ultimately paved the way for a successful implementation.

The pilot project in Canada served as a proving ground, enabling us to test our assumptions and methodologies in a less pressured environment. By targeting a different segment and concentrating on a specific aspect of the supply chain, we built trust and demonstrated the value of predictive models.

Engaging with executives in similar roles but in different parts of the business enabled us to gain a clearer picture of how decisions were made and the flow of information. By involving multiple stakeholders from the outset and fostering an inclusive environment, we reduced information asymmetries and kept everyone engaged and aligned. As the internal teams invested more time and effort into the project, their commitment grew, driven by the desire to see the project succeed. This dynamic, combined with the positive results we were able to demonstrate, created a reinforcing cycle of commitment and enthusiasm, ultimately driving the project forward successfully.

This experience underscored the need for patience, empathy, and a flexible, adaptive approach when implementing innovative solutions. By aligning technical expertise with organizational dynamics and maintaining open communication, we created a robust framework for ongoing collaboration and improvement. This journey reinforced the importance of addressing cognitive biases head-on and the value of fostering a collaborative, transparent environment to achieve successful outcomes.

The implementation of predictive models and advanced analytics in an organizational setting is not just a technical challenge but a human one. Understanding and addressing cognitive biases like loss aversion and status quo bias are essential for driving change. It's crucial to build trust through small, low-risk pilots, and to involve all stakeholders from the beginning. Open communication, frequent touchpoints, and a collaborative spirit are key to overcoming resistance and achieving success. This journey has shown that the intersection of technical expertise and behavioral economics can unlock tremendous value, leading to better decision-making, improved customer experiences, and, ultimately, a more resilient and responsive organization.

CHAPTER

7

Unlocking Scale
Overcoming Operational and Organizational Complexity in Scaling AI Projects

"Simplicity is the ultimate sophistication."
– Leonardo da Vinci

"Complexity is your enemy. Any fool can make something complicated. It is hard to make something simple."
– Richard Branson

Navigating the operational and organizational complexities inherent in scaling artificial intelligence (AI) projects is no small feat. The examples in the previous chapter underscore the critical role that behavioral economics can play in facilitating effective change management. By understanding and addressing cognitive biases, fostering collaboration, and ensuring clear communication, organizations can overcome many of the barriers that hinder the successful deployment of AI initiatives. This chapter delves into the practical lessons learned and the key enablers for achieving scalability in AI and machine learning (ML) projects.

Enablers of Success

I learned many lessons in the stories that I shared previously. As practitioners, it is sometimes hard to remember, as obvious as it might be, that there is a large trust and information gap between different groups to ensure the success and scalability of AI and ML projects. There are three main enablers that I have come to appreciate in this process. Anytime any of these three components is missing, something goes wrong.

The first enabler is infrastructure: Investments in modern data, AI, and ML infrastructure are essential to ensure scalability and realizing true value. Without a solid technological foundation, even the most well-conceived AI strategies can falter. This includes robust data pipelines, scalable computing resources, and advanced analytical tools.

The second enabler is building capacity and team capabilities: The right team is critical for long-term success. This team should be interdisciplinary, drawing expertise from business, data science, engineering, and behavioral sciences. Intellectual diversity matters because it brings different perspectives and problem-solving approaches, ensuring a more holistic understanding and addressing the multifaceted challenges of AI projects.

The third enabler is the operational model and close collaboration among different teams: Ensuring alignment with business strategy, evaluating gaps, and addressing risks is vital. Effective collaboration and communication frameworks need to be in place to bridge the gap between technical teams and business stakeholders. This alignment helps in setting realistic expectations, defining clear goals, and maintaining a shared vision throughout the project's life cycle.

By focusing on these enablers and applying principles from behavioral economics, organizations can better manage the complexities associated with scaling AI projects. This approach not only fosters a culture of trust and collaboration but also ensures that technical advancements are seamlessly integrated into the broader business strategy, paving the way for sustained success.

I won't dwell too much on the first enabler because much of the conversation about enterprise AI already centers on topics such as cloud infrastructure, data operations (DataOps), model operations (ModelOps), and data and model governance. Major conferences by industry leaders like Amazon Web Services, Microsoft, Gartner, and Google dedicate substantial time to these crucial topics. These events offer deep dives into best practices, emerging trends, and the latest technologies that form the backbone of scalable AI infrastructures. Investments in modern data and AI infrastructure are undeniably vital. This infrastructure includes scalable cloud platforms that can handle vast amounts of data, sophisticated DataOps practices to streamline data management, and ModelOps to ensure smooth deployment and monitoring of AI models. Effective data and model governance frameworks are also essential to maintain the integrity, security, and compliance of AI systems.

The real challenge – and the area where organizations often stumble – lies in the human and organizational elements of AI implementation. This brings us to the second and third enablers: building team capacity and fostering effective collaboration.

Developing deep technologies such as AI is scientifically challenging and requires years of research, experimentation, and trial and error. Implementing and scaling these technologies in complex environments like multinational enterprises is even more complicated. Information and incentive structures often hinder the effective augmentation of human systems, making it difficult to identify real gaps and increase efficiencies.

One major issue is that the end users of these analytics and predictive systems are often not included in the design process or are brought in too late. The complexities and subtleties of their roles, as well as the decisions they are responsible for, are frequently overlooked. This disconnect can lead to solutions that are technically sound but fail to address the practical needs and challenges faced by these end users.

Furthermore, the IT organization, despite having the best intentions, might procure out-of-the-box technologies that promise increased efficiencies and savings. However, one size does not fit all. The availability of a tool does not guarantee its adoption and use either. For AI solutions to be effective, they need to be tailored to the specific context of the organization and the unique needs of its users.

That's why intentionally building teams that can bridge the gaps between different parts of the organization is as important, if not more, than developing the right solutions and high-quality models. These teams play essential roles in transmitting information and shedding light on important nuances that might point to biases or gaps in the data, overlooked key performance indicators, or business constraints that could affect the implementation and scalability stages of a project. We've all heard of or seen projects in which, after several months of development or implementation, an executive innocently asks during a presentation, "What about X? Have you considered Y?" The development team then looks at each other as if they've seen a ghost because they've never heard of X or Y.

An inclusive design process involves engaging end users early in the design process to understand their workflows, pain points, and decision-making processes. This ensures that the solutions developed are aligned with their needs and are more likely to be adopted. Maintaining transparency and communication throughout the project is crucial; regularly updating stakeholders on progress, challenges, and successes builds trust and reduces resistance to change. Providing comprehensive training and education helps users understand and effectively use the new technologies, alleviating fears and building confidence in the new systems. Implementing an iterative approach to development and deployment, starting with small pilots (remember minimum viable data?!), gathering feedback, and making necessary adjustments before scaling up, allows for continuous improvement and ensures the final solution is well tuned to users' needs. Aligning incentives with desired outcomes by recognizing and rewarding those who contribute to the successful implementation and adoption of new technologies can motivate individuals to embrace change and actively participate in the process. Finally, leveraging insights from behavioral economics to design interventions that address cognitive biases, such as framing changes positively, using social proof to show peer adoption, and setting default options that favor desired behaviors, can significantly influence adoption rates.

By focusing on these strategies, organizations can navigate the complexities of implementing and scaling AI technologies. Combining a robust technological foundation with a deep understanding of human behavior and organizational dynamics can lead to more successful and sustainable AI initiatives. The lessons learned from previous projects highlight the importance of inclusive design, transparent communication, and aligning incentives to drive meaningful change.

Communication and Intellectual Diversity

Effective communication and intellectual diversity are crucial for the success of multidisciplinary teams, especially in the realm of

AI and data analytics. Bridging the gap between different technical backgrounds and business stakeholders is vital for creating innovative solutions and ensuring their successful implementation. However, this task is often challenging due to the inherent differences in language, jargon, and problem-solving approaches across various disciplines.

Something that has often made me laugh is when many executives, especially senior ones, tell me, "You understand us!" as if it is such a unique phenomenon. What surprises them is that someone with my technical background speaks their language, complete with their jargon and acronyms. Though, once, a senior executive pulled me aside and told me that I shouldn't go around the organization talking about "experimentation." He said I was scaring people and suggested that I use "test and learn" instead.

Before discussing what a multidisciplinary team across the organization should look like, let's talk about what a diverse data and analytics team should look like. A few years ago, I was invited by a friend to speak at a special program that brought together graduate students from different disciplines such as physics, computer science, economics, psychology, and sociology at the University of Virginia. I shared funny stories with the class about how I often watch with amusement the interesting and passionate conversations among economists, physicists, and computer scientists on my team. I also mentioned that our team was composed of 50% men and 50% women, from 11 different countries, speaking 18 different languages.

During the session, one of the students raised her hand and asked how I help the team when they have "problems" with communications. This question piqued my curiosity because I never framed communication among my team members as a problem or challenge. So, I asked her why she thought there was a problem. She responded that it must be difficult for people from different disciplines to work together, especially on high-stakes projects. I then asked which part of communication she thought was more difficult. After all, communication is composed of two parts: listening and speaking. She wasn't sure how to answer.

I explained that transmitting our knowledge and information can indeed be difficult, especially to other groups unfamiliar with our jargon. However, with practice and patience, anyone can learn and improve their speaking skills. But what about the listening part? When someone from a different field is speaking to us and struggling to communicate complex topics from their discipline, how well and patiently do we listen? In my experience, this is the harder part – not just because they are using a language we might not be familiar with, but because we often believe that our own discipline and expertise offer the best method for solving the problem at hand. Why would we spend so much time specializing in a field if we didn't believe that? So, most of the time, we are merely politely waiting for others to finish so that we can show them how our way is better.

As you can imagine, the room was silent. Pointing to the elephant in the room lifted a certain degree of tension, and the conversation became much more fluid. Suddenly, there was a flood of questions about what they could do better and how they could collaborate more effectively. My answer made them laugh in its simplicity: BE CURIOUS. Be very curious.

This curiosity has been the key trait that has made my team successful and effective. Many executives I work with often ask me how I have managed to have so many PhDs work on practical problems. They are right – a lot of researchers have a hard time transitioning out of academia. But the remarkable thing is getting so many people with mixed backgrounds and disciplines to work together and collaborate with our clients. What makes it work is our genuine curiosity about different solutions, various ways of looking at the question at hand, our clients, the complexity of their environment, and more. This curiosity makes us humble and more patient.

Diversity matters, but in this field, intellectual diversity is even more essential. The right composition of scientists and engineers is what truly differentiates innovative teams that will supercharge the organization from those who will follow the status quo because change is difficult. We have worked with some data and AI/ML teams that were as homogeneous as it gets – everyone had a graduate degree in computer science. To their executives, we were very "creative"! The truth is that our toolkit is simply more diverse than their internal teams, making it easier to mix and match.

By fostering an environment where active listening is prioritized, and where team members are encouraged to truly understand each other's perspectives, we can bridge the communication gap that often exists in multidisciplinary teams. This approach not only enhances collaboration but also drives innovation because it allows for the integration of diverse viewpoints and expertise. When team members feel heard and valued, they are more likely to contribute their unique insights and work together toward common goals, ultimately leading to more effective and scalable AI solutions.

The same principles apply to collaboration across different teams within an enterprise. Just as multidisciplinary teams benefit from intellectual diversity and curiosity, so too do cross-functional teams within a company. When different departments – such as marketing, operations, finance, and IT – collaborate, they bring unique perspectives and expertise to the table. However, these teams often face communication barriers due to differences in jargon, priorities, and methodologies.

To bridge these gaps, fostering a culture of curiosity is crucial. Encouraging team members to be genuinely curious about each other's work and to ask questions can lead to a deeper understanding and more effective collaboration. This curiosity helps break down silos and promotes a more integrated approach to problem-solving.

I remember one of our early projects, just before a workshop we organized with the senior team members in the customer experience and marketing department. I asked my main stakeholder if there were some other "irrelevant" people we could also invite, such as the chief product officer (CPO) or chief revenue officer (CRO) and their teams. He laughed, understanding I meant those not directly involved in the project. Though hesitant, he agreed to extend the invitation. To his surprise, everyone accepted, making him quite nervous. He warned me about the CPO's impatience and directness. I smiled, knowing it would be a fun group to manage. What was initially a workshop for 4 people turned into a session with 15, including a couple of C-suite executives.

When the CPO joined the meeting, the whole room was tense. The game theorist in me was fascinated; I felt like a kid in a candy store. He informed me he would only stay for 15 minutes, so we got started. Instead of diving into our project and goals, I shared my background and expressed my deep interest in their toughest challenges as they embarked on a steep growth trajectory. They began discussing issues they believed were unique to their teams, and I subtly connected the dots, showing how something the CPO mentioned was directly related to a problem the CPO was interested in, and so on.

The CPO ended up staying the entire four hours. We had a brilliant conversation, leading to a much deeper appreciation of the interdependencies of their challenges and pain points. The next day, I even received a surprise visit from the CRO, apologizing for missing the conversation.

Since then, I always ask my stakeholders to invite "irrelevant" people to the room, and I chuckle as I say it. This approach consistently proves valuable, fostering richer discussions and uncovering connections that might otherwise go unnoticed. By engaging a diverse group, we ensure a more holistic understanding of the issues at hand and pave the way for more integrated and effective solutions.

Furthermore, creating opportunities for regular interaction and knowledge sharing between departments can enhance this collaborative spirit. Joint workshops, cross-departmental projects, and shared goals can help teams see the bigger picture and understand how their work affects the organization as a whole.

In addition, promoting active listening within the organization can significantly improve collaboration. When team members feel heard and understood, they are more likely to engage fully and contribute their best ideas. This active listening should be practiced at all levels of the organization, from executives to frontline employees.

Ultimately, the principles of intellectual diversity, curiosity, and active listening are foundational to building effective, innovative, and collaborative teams – both within specialized data science units and across the broader enterprise. By embracing these principles, organizations can overcome communication barriers, leverage diverse perspectives, and drive meaningful change.

Building Trust, Experimentation, and Adoption

A few years ago, I received an intriguing question from one of my stakeholders. He asked if there was a way to transform black box models into more transparent, clear box ones. This made me pause because I pondered what he truly wanted to achieve. Was it to make the models interpretable? But interpretable for whom? For technical data science teams to help monitor performance and detect bias or data drift? Or for business stakeholders and end users who were supposed to use the insights from these predictive models in their decision-making processes, thus promoting trust and increasing adoption?

The former is much easier for data scientists than the latter. Data scientists are accustomed to, and indeed often have, multiple models, processes, and unit tests throughout the data and model pipelines to monitor performance, look for anomalies and bias, and identify suspicious changes. These measures ensure that we can trust the predictions and outcomes of the models. Additionally, we perform extensive statistical analysis on the raw data to familiarize ourselves with trends and patterns and gain a better understanding of underlying factors that might lead to important predictive signals. In economics, we call this *developing intuition*.

However, making models interpretable for business stakeholders and end users is a different challenge. These users need to understand not just the outputs of the models, but the reasoning behind these outputs to make informed decisions. This requires translating complex statistical concepts into intuitive insights that align with their business context. It's about bridging the gap between technical accuracy and practical relevance.

Business stakeholders typically have two types of questions. The first type concerns the effectiveness of the models and attribution: Are the recommendations from the models effective? What kind of a difference do they create? How can we correctly attribute the changes we observe to the model if we incorporate its insights? These questions are based on understanding the tangible impact of the model and ensuring that its predictions translate into measurable outcomes.

The second type of question relates to risk and the underlying mechanics of the models: How do they work, and why do they produce the results they do? This is especially pertinent when the risk of making a mistake or missing something is high, or when the results are counterintuitive or don't align with their expectations. Stakeholders need to understand the reasoning behind the model's predictions to trust and act on them confidently.

Consider a scenario in a retail company where a predictive model forecasts sales for the next quarter. For the data science team, it's relatively straightforward to validate the model's accuracy through various statistical tests and performance metrics. They can delve into the nuances of feature importance, model coefficients, and residual analysis. However, for the business stakeholders – like the chief marketing officer or CRO – these technical details might be overwhelming and impractical for strategic decision-making.

Addressing these concerns requires a strategy that involves both rigorous validation and transparent communication. It's about demonstrating the effectiveness of the models through clear, quantifiable metrics and providing a narrative that explains how the models work and why they generate certain outcomes. By doing so, we can bridge the gap between technical sophistication and practical application, ensuring that business stakeholders feel confident and informed in their decision-making processes.

Interpretation Layers

When my stakeholder, who led the data and analytics efforts in North America for one of the largest consumer packaged goods companies in the world, asked me the question about turning the black box into a clear box, he had both scenarios in mind: creating clarity and transparency for both the technical user as well as the business stakeholder.

We were developing ML models for forecasting demand and optimizing pricing and promotions for their products. The outcomes from these models included recommended prices and discount levels for various products over the next 52 weeks. These results aimed to assist account executives in negotiating with grocers and other retailers and improve promotion efficiencies.

This activity is typically governed by a collaborative joint business planning process between a CPG company and a grocer or retailer. This process involves creating a yearly plan (52-week promo calendar) and making in-year adjustments to individual promotions based on factors such as retailer constraints, budgeting, and pricing activities. Promo spend efficiency is a key objective of joint business planning because inefficiencies can lead to investment losses and missed opportunities. Moreover, inefficient promo spending directly affects the supplier's ability to provide the right product at the right time, causing undesired effects along the supply chain.

Therefore, ensuring that the ML models were not only accurate but also trusted and understood by the business stakeholders was critical for achieving these objectives. This required clear communication of how the models worked, their effectiveness, and how their recommendations could be confidently acted on to drive business value.

We were working with both their internal data science team and the business stakeholders, who were account executives working directly with their customers – grocers and retailers.

Because these models were related to a core business function, it was crucial for his internal team to thoroughly vet the models and have complete trust in their process and performance. Simultaneously, it was essential to equip his business counterparts with actionable insights that they trusted and could use effectively in negotiations.

To achieve this, we incorporated interpretation layers into the predictive models to make them more transparent to the end users. For the data science team, these interpretation layers included detailed performance metrics, model diagnostics, and validation results. This transparency enabled the technical team to monitor the models, detect biases, and ensure that the predictions remained reliable over time. We also provided tools for them to perform their own analyses and validations, reinforcing their confidence in the models.

To design the interpretation layer for the business stakeholders, we first needed to understand the core metrics and factors they use to sanity check the outcomes. Additionally, we had to anticipate the questions their counterparts might ask them and ensure they were equipped with the necessary answers and insights. We held several working sessions with our stakeholders, and the following questions emerged: What is the impact of the base price? What is the effect of the promotion? How many additional units will we sell because of this promotion? Does this discount cannibalize our own products? What is the balance between the additional revenue and profit? How do I know this is the best promotion? How do the outcomes compare with the promotion we ran last year? Why aren't we using the same promotion as last year?

It's easy to see traces of loss aversion and status quo bias in these questions. They were also anchoring their expectations on the "known" outcomes from last year, highlighting the combined influence of anchoring and availability heuristics. These cognitive biases underscore the importance of the interpretation layer in not only providing clear answers but also contextualizing the data to address these inherent biases. By doing so, we can help the stakeholders move beyond their reliance on past practices and embrace more data-driven decision-making.

So, we decided to create a simple interface that directly addressed these questions. For each price and recommended promotion, we highlighted the impact of the base price followed by the promotional price. Specifically, we showed how many units would be sold as a result of the base price without any promotions, and how many units would be sold as a result of the discount. This was followed by the expected revenue and profit, with a clear indication of the effect of promotion fatigue.

Next, we presented the historical performance of the product over a customizable period, such as six months. We highlighted the periods when a promotion was present, separating the impact of the base price from the promotional price. We also provided an aggregated view showing the total revenue, profit, and units sold.

Additionally, we recommended incorporating a simulation engine that would enable users to answer what-if questions, such as the impact of increasing prices by 5% or exploring other scenarios. This feature enabled them to "play" with the model, generating simulations and comparing the outcomes with their knowledge and past experiences.

This comprehensive approach not only provided clarity and transparency but also helped increase their conviction in the outcomes. By making the predictive models more interactive and relatable to their day-to-day decision-making processes, we fostered greater trust and adoption among the business stakeholders.

The Power of Experimentation

In the previous chapter, we delved into the organizational complexities and challenges associated with scaling AI projects within enterprises. We explored how behavioral economics and effective change management practices can help bridge the gap between different technical backgrounds and business stakeholders, fostering a collaborative environment conducive to innovation. These approaches are all important, but they don't make the ML and AI models easier to understand, beyond highlighting the important correlations, or fail to explain the underlying causal mechanisms.

This is where methodologies from econometrics and behavioral economics come in handy again and enable us to move beyond mere correlations and uncover the deeper connections between factors that drive behavior and value. Understanding these causal relationships is crucial for making informed decisions, optimizing strategies, and ultimately driving business value.

Economics provides a robust framework for moving beyond correlations and exploring causal relationships. Techniques such as regression analysis, instrumental variables, and difference-in-differences approaches are designed to control for confounding variables and identify causation. These methods help isolate the effect of one variable on another, providing more reliable insights into how different factors drive outcomes.

For example, regression analysis can control for various factors that might influence sales, such as market conditions and advertising channels, to determine the true impact of marketing spend on sales. Similarly, the instrumental variables approach can address endogeneity issues, where independent variables are correlated with the error term, ensuring a more accurate estimation of causal effects.

Although advanced analytical methods can provide significant insights, experimentation remains the gold standard for establishing causality. Controlled experiments, such as A/B testing and RCTs (randomized control trials), enable us to manipulate variables and observe their effects in a controlled setting. By randomly assigning subjects to different groups and comparing outcomes, we can determine whether a particular intervention causes a specific effect. Experimentation helps us isolate variables and determine their true effects, providing robust evidence of what works and what doesn't. This empirical approach not only enhances our understanding of causal mechanisms but also enables us to measure the return on investment (ROI) of various strategies and initiatives. By moving beyond correlations and embracing a more sophisticated analytical framework, we can harness the full potential of AI and data science to generate meaningful, actionable insights and drive more effective decision-making.

Measuring ROI Through Experimentation

Experimentation also enables us to measure the ROI of different strategies and interventions. By quantifying the impact of specific actions, we can make more informed decisions about where to allocate resources. For example, a marketing team might use RCTs to evaluate the effectiveness of different advertising campaigns, measuring the incremental sales generated by each campaign and calculating the ROI. This empirical approach ensures that resources are directed toward initiatives that deliver the highest returns, optimizing the overall strategy and maximizing value creation.

Let me give you an example of one of our projects from a few years ago. We worked with a credit union that used their call center agents to contact members whenever they wanted to promote a new financial product, such as a new term deposit campaign. In previous chapters, I discussed how we combined behavioral segmentation with propensity models to build targeted campaigns. The models were performing well, but we were unsure about the impact of the intervention through call centers.

We decided to run three different experiments. The first one was a test to get the call center agents familiar and comfortable with the concept. The next two were designed to measure the impact of the personalized calls to a select list of members.

For these campaigns, we divided the list of customers that had been prioritized using the ML models into a control group (no contact) and a test group (contacted customers). This method enabled us to establish the direct impact of the intervention.

The results of the experimentation were impressive. Two term deposit campaigns were executed using these lists, yielding significant improvements: The spring campaign generated $30.2 million in total over eight weeks, increasing the number of deposits by 53.75% and the amount committed to term deposits by 50.48% compared to the control group. The fall campaign generated $53.9 million in total over four weeks, increasing the number of deposits by 46% and the amount committed to term deposits by 45% compared to the control group.

The improvement from one campaign to the next was due to the retraining feedback loop built into the ML models, which continuously strengthened insights and outputs. The targeted lists rendered outreach more intentional and positively affected customer engagement.

The success of these campaigns highlighted the power of predictive models and the value of experimentation in establishing causality and measuring ROI. By using a structured, data-driven approach, the organization was able to optimize its marketing efforts and improve customer engagement. This case exemplifies how rigorous experimentation, combined with advanced analytics, can drive substantial business outcomes and foster a culture of data-driven decision-making. The targeted, empirical approach not only proved the effectiveness of the interventions but also built a foundation for continuous improvement and strategic resource allocation, ultimately enhancing overall business performance and customer satisfaction.

Conclusion

I hope the journey through this chapter has highlighted some of the critical elements necessary for the successful implementation and scaling of AI projects in complex organizational environments. From understanding and overcoming cognitive biases to fostering intellectual diversity and effective communication, each aspect plays a pivotal role in bridging the gap between different technical backgrounds and business stakeholders.

We've seen how a robust infrastructure, combined with the right team capabilities and an operational model that promotes collaboration, can set the stage for AI success. The inclusion of behavioral economics principles helps manage change more effectively, ensuring that all stakeholders are aligned and engaged.

The stories and examples provided illustrate the importance of curiosity, active listening, and inclusive design processes. By involving end users early, maintaining transparency, providing adequate training, and iteratively testing and refining models, organizations can build trust and ensure the practical application of AI technologies. Aligning incentives and leveraging behavioral insights further drive the adoption and effective use of these innovations.

As we transition to exploring deeper connections and establishing causal relationships, the incorporation of interpretation layers and rigorous experimentation becomes indispensable. Interpretation layers serve as the bridge between complex model outputs and actionable business insights. For data scientists, these layers offer detailed performance metrics and tools for continuous monitoring, ensuring reliability and unbiased predictions. For business stakeholders, they translate technical jargon into comprehensible explanations, building confidence and facilitating the integration of AI-driven insights into decision-making processes.

Experimentation, particularly through randomized controlled trials, provides the empirical backbone of AI initiatives. It enables us to isolate variables, determine true impacts, and quantify benefits. This iterative approach, with its feedback loops and continuous improvement mechanisms, ensures that models and strategies evolve based on real-world performance, further enhancing their effectiveness and reliability.

Together, interpretation layers and experimentation create a powerful framework for driving AI adoption and trust. By demystifying the black box of AI and providing empirical evidence for its interventions, these tools enhance credibility and empower organizations to make data-driven decisions with confidence.

The path to AI success is multifaceted, requiring a balance of technical prowess, human understanding, and strategic alignment. As we move forward, the lessons learned from these experiences will guide us in creating more integrated, responsive, and innovative AI solutions. By fostering an environment where diverse perspectives are valued and collaboration is prioritized, we can navigate the complexities of AI implementation and unlock its full potential for transformative impact.

Epilogue

"Everything must be made as simple as possible. But not simpler."

– Albert Einstein

A few years ago, a dear friend of mine published a book. When we met shortly after, I asked him about his experience writing the book. He shared how he learned more about himself through the process. This was certainly true for me as well. Reflecting on many stories, conversations, and hard, risky decisions reminded me that being at the forefront of innovation requires strong conviction, hard work, and an unwavering commitment to excellence.

Coincidentally, in the last few months, while working on this manuscript, I have been reading biographies and autobiographies of incredible scientists such as Richard Feynman, Albert Einstein, Jennifer Doudna, and J. Robert Oppenheimer. I also delved into *The Geek Way* (McAfee, 2023) and *The Innovator's Dilemma* (Christensen, 2003). The parallels between the scientific methodologies and the struggles these remarkable individuals experienced to advance science and technology, and what many "geeks" in organizations and companies are doing to push the boundaries of what is possible, are truly striking.

Writing this book has been not only an exercise in documenting and sharing insights but also a journey of self-discovery. The act of distilling complex ideas and experiences into a narrative that others can understand and learn from has made me revisit the core principles that have driven my work. I have been reminded time and again that the path of innovation is fraught with challenges, but it is also incredibly rewarding.

Throughout these chapters, I've shared the stories of big data and AI projects, illustrating how economic intuition and behavioral economic methods can be powerful tools in decoding complex patterns and driving impactful decisions. From transforming customer engagement strategies to optimizing supply chains, the applications of these interdisciplinary approaches are vast and varied.

As I look back on my journey, I am filled with gratitude for the opportunities to work with brilliant minds across different fields and industries. I am indebted to all who believed in me, cheered me on, and supported me and my crazy ideas. Thank you for your patience, for trusting the process and your unwavering support. I dedicate this book to my incredible team. Their curiosity, resilience, and dedication have been a constant source of inspiration. Together, we have tackled seemingly insurmountable problems, pushing the limits of what is possible and redefining the boundaries of innovation.

The stories and lessons in this book are not just about technology and data; they are about people – their behaviors, motivations, and the intricate dance between intuition and analytics. It is about understanding the human element in every equation and using that understanding to create solutions that are not only effective but also resonate on a deeper level. As we stand on the brink of an era when AI and machine learning are poised to revolutionize every aspect of our lives, I hope the insights shared in this book will inspire you to embrace these tools with a sense of

Epilogue

curiosity and purpose. Let us continue to ask the right questions, be curious, and above all, remain committed to the relentless pursuit of excellence.

Thank you for joining me on this journey. Here's to the innovators, the thinkers, and the dreamers – may we always strive to turn the impossible into reality.

Notes

Preface

1. Safi Bahcall, *Loonshots: How to Nurture the Crazy Ideas That Win Wars, Cure Diseases, and Transform Industries* (New York: St. Martin's Press, 2019).
2. Matthew Jackson, *The Social and Economic Networks* (Princeton, NJ: Princeton University Press, 2008).

Chapter 1

1. In April 2023, industry reports typically cited an average online shopping cart abandonment rate of about 60%–80%. Baymard Institute, "Average Cart Abandonment Rate," April 2023, https://baymard.com/lists/cart-abandonment-rate.
2. Dan Ariely, *Predictably Irrational: The Hidden Forces that Shape our Decisions* (London, UK: Harper Collins, 2008).
3. Mary Loxton, Robert Truskett, Brigitte Scarf, et al., "Consumer Behavior During Crises: Preliminary Research on How Coronavirus Has Manifested Consumer Panic Buying, Herd Mentality, Changing Discretionary Spending and the Role of the Media in Influencing Behaviour," *Journal of Risk and Financial Management* 13, no. 8 (2020): 166, https://www.mdpi.com/1911-8074/13/8/166.
4. Arpana Sharma, Madhu Pruthi, and Geetanjali Sageena, "Adoption of Telehealth Technologies: An Approach to improving Healthcare System," *Translational Medicine Communications* 7, no. 1 (2022): 20, 10.1186/s41231-022-00125-5.

Chapter 2

1. Richard H. Thaler, "From Homo Economicus to Homo Sapiens," *Journal of Economic Perspectives* 14, no. 1 (2000): 133–41, 10.1257/jep.14.1.133.
2. Thorstein Veblen and Martha Banta, *The Theory of the Leisure Class* (Oxford, UK: Oxford University Press, 2009).
3. John Maynard Keynes, *The General Theory of Employment, Interest and Money* (London: Palgrave Macmillan, 1936).
4. Herbert A. Simon and Robert Dahl, "Administrative Behavior: A Study of Decision-Making Processes in Administrative Organization," *Administrative Science Quarterly* 2, no. 2 (1957): 244–48, 10.2307/2390693.
5. George Katona, *Psychological Analysis of Economic Behaviour* (Westport, CT: Greenwood Press, 1977).
6. Amos Tversky and Daniel Kahneman, "Judgment Under Uncertainty: Heuristics and Biases," *Judgment Under Uncertainty*, ed. Daniel Kahneman, Paul Slovic, and Amos Tversky (Cambridge, UK: Cambridge University Press, 1982), 3–20.
7. Daniel Kahneman and Amos Tversky, "Prospect Theory: An Analysis of Decision Under Risk," *Econometrica* 47, no. 2 (1979): 263, 10.2307/1914185.
8. Kahneman and Tversky, "*Prospect Theory*," 263.
9. Richard H. Thaler and Cass R. Sunstein, *Nudge: Improving Decisions About Health, Wealth, and Happiness* (New York: Penguin Books, 2009).
10. Thaler and Sunstein, *Nudge*.
11. Jill Rutter, "'Nudge Unit'," Institute for Government, March 2, 2010, www.instituteforgovernment.org.uk/article/explainer/nudge-unit.
12. Richard H. Thaler, "Mental Accounting and Consumer Choice," *Marketing Science* 27, no. 1 (2008): 15–25, 10.1287/mksc.1070.0330.
13. Richard H. Thaler and Shlomo Benartzi, "Save More Tomorrow™: Using Behavioral Economics to Increase Employee Saving," *Journal of Political Economy* 112, no. S1 (2004), 10.1086/380085.
14. Richard H. Thaler, "The End of Behavioral Finance," *Financial Analysts Journal* 55, no. 6 (1999): 12–17, 10.2469/faj.v55.n6.2310.
15. Prerna Juneja and Tanushree Mitra, "Auditing E-Commerce Platforms for Algorithmically Curated Vaccine Misinformation," *Proceedings of the 2021 CHI Conference on Human Factors in Computing Systems*, May 6, 2021, 10.1145/3411764.3445250.
16. Alan R. Dennis, Lingyao Yuan, Xuan Feng, et al., "Digital Nudging: Numeric and Semantic Priming in e-Commerce," *Journal of Management Information Systems* 37, no. 1 (2020): 39–65, 10.1080/07421222.2019.1705505.
17. Dan Ariely, *Predictably Irrational: The Hidden Forces That Shape Our Decisions* (London: Harper Collins, 2008).

18. Ariely, *Predictably Irrational*.
19. Ariely, *Predictably Irrational*.
20. Ratnalekha V. N. Viswanadham, "How Behavioral Economics Can Inform The Next Mass Vaccination Campaign: A Narrative Review," *Preventive Medicine Reports* 32 (2023): 102118, 10.1016/j.pmedr.2023.102118.
21. Daniel Guinart and John M. Kane, "Use of Behavioral Economics to Improve Medication Adherence in Severe Mental Illness," *Psychiatric Services* 70, no. 10 (2019): 955–57, 10.1176/appi.ps.201900116.
22. Oana M Blaga, Livia Vasilescu, and Razvan M Chereches, "Use and Effectiveness of Behavioural Economics in Interventions for Lifestyle Risk Factors of Non-Communicable Diseases: A Systematic Review with Policy Implications," *Perspectives in* Public Health 138, no. 2 (2017): 100–10, 10.1177/1757913917720233.
23. Benjamin D. Horne, Joseph B. Muhlestein, Donald L. Lappé, et al., "Behavioral Nudges as Patient Decision Support For Medication Adherence: The Encourage Randomized Controlled Trial," *American Heart Journal* 244 (2022): 125–34, 10.1016/j.ahj.2021.11.001.
24. Emily Herrett, Tjeerd van Staa, Caroline Free, et al., "Text Messaging Reminders for Influenza Vaccine In Primary Care: Protocol for a Cluster Randomised Controlled Trial (TXT4FLUJAB)," *BMJ Open* 4, no. 5 (2014), 10.1136/bmjopen-2013-004633.
25. Leslie L. Chang, Adam D. DeVore, Bradi B. Grander, et al., "Leveraging Behavioral Economics to Improve Heart Failure Care and Outcomes," *Circulation* 136, no. 8 (2017): 765–72, 10.1161/circulationaha.117.028380.
26. Dan Ariely, "Will Pay for Performance Backfire? Insights from Behavioral Economics," *Forefront Group*, October 11, 2012, 10.1377/forefront.20121011.023909.
27. Leslie L. Chang, Adam D. DeVore, Bradi B. Grander, et al., "Leveraging Behavioral Economics."
28. Dan Ariely, "Combining Experiences over Time: The Effects of Duration, Intensity Changes and On-Line Measurements on retrospective Pain Evaluations," *Journal of Behavioral Decision Making* 11, no. 1 (1998): 19–45, 10.1002/(sici)1099-0771(199803)11:1<19::aid-bdm277>3.0.co;2-b.
29. Dan Ariely and Klaus Wertenbroch, "Procrastination, Deadlines, and Performance: Self-Control by Precommitment," *Psychological Science* 13, no. 3 (2002): 219–24, 10.1111/1467-9280.00441.
30. Dan Ariely, "Combining Experiences over Time," 19–45.
31. Gustav Jahoda, "A Cross-Cultural Perspective on Developmental Psychology," *International Journal of Behavioral Development* 9, no. 4 (1986): 417–37, 10.1177/016502548600900402.
32. Devkant Kala, Dhani Shanker Chaubey, and Ahmad Samed Al-Adwan, "Cryptocurrency Investment Behaviour of Young Indians: Mediating

Role of Fear of Missing Out," *Global Knowledge, Memory and Communication* (2023), 10.1108/gkmc-07-2023-0237.
33. Richard L. Peterson, "The Neuroscience of Investing: fMRI of the Reward System," *Brain Research Bulletin* 67, no. 5 (2005): 391–97, 10.1016/j.brainresbull.2005.06.015.
34. James Heyman and Dan Ariely, "Effort for Payment," *Psychological Science* 15, no. 11 (2004): 787–93, 10.1111/j.0956-7976.2004.00757.x.
35. Chang-Yuan Lee, Carey K. Morewedge, Guy Hochman, et al., "Small Probabilistic Discounts Stimulate Spending: Pain of Paying in Price Promotions," *Journal of the Association for Consumer Research* 4, no. 2 (2019): 160–71, 10.1086/701901.
36. Ekaterina Smironva, Kiattipoom Kiatkawsin, Seul Ki Lee, et al., "Self-Selection and Non-Response Biases in Customers' Hotel Ratings – A Comparison of Online and Offline Ratings," *Current Issues in Tourism* 23, no. 10 (2019): 1191–1204, 10.1080/13683500.2019.1599828.
37. Daniel Kahneman, *Thinking Fast and Slow* (London: Penguin Books, 2011).
38. Amos Tversky and Daniel Kahneman, *Judgment Under Uncertainty*, 3–20.
39. Daniel Kahneman, Olivier Sibony, and Cass R. Sunstein, *Noise: A Flaw in Human Judgment* (New York: Little, Brown Spark, 2022).
40. Kahneman and Tversky, *"Prospect Theory,"* 263.
41. Richard Thaler, Cass R. Sunstein, and John P. Balz. "Choice Architecture," *The Behavioral Foundations of Public Policy*, ed. Eldar Shafir (Princeton, NJ: Princeton University Press, 2013), 428–39.
42. Thaler and Benartzi, "Save More Tomorrow™."
43. Robert Kolb, "Pareto, Vilfredo (1848–1923)," *Encyclopedia of Business Ethics and Society* (Thousand Oaks, CA: SAGE Publications, 2008).
44. P. A. Sloan and Irving Fisher, "The Theory of Interest," *The Economic Journal* 41, no. 161 (1931): 84, 10.2307/2224140.
45. Edward H. Chamberlin, "An Experimental Imperfect Market," *Journal of Political Economy* 56, no. 2 (1948): 95–108, 10.1086/256654.
46. Vernon L. Smith, *Economics in the Laboratory* (Tucson, AZ: American Economic Association, 1994).
47. Francesca Gino, "Uber Shows How Not to Apply Behavioral Economics," *Harvard Business Review*, April 13, 2017, https://hbr.org/2017/04/uber-shows-how-not-to-apply-behavioral-economics.
48. Noam Scheiber, "How Uber Uses Psychological Tricks to Push Its Drivers' Buttons," *New York Times*, April 2, 2017, https://www.nytimes.com/interactive/2017/04/02/technology/uber-drivers-psychological-tricks.html.
49. Dan Ariely, "Column: Why Businesses Do Not Experiment," *Harvard Business Review*, August 1, 2014, https://hbr.org/2010/04/column-why-businesses-dont-experiment.

Chapter 3

1. ATLAS is one of two general-purpose detectors that will stand in the LHC ring and is being assembled in its underground cavern. Construction of the ATLAS detector was completed in 2008 and the experiment detected its first single proton beam events on 10 September of that year. Data-taking was then interrupted for over a year due to an LHC magnet quench incident. On 23 November 2009, the first proton–proton collisions occurred at the LHC and were recorded by ATLAS, at a relatively low injection energy of 900 GeV in the center of mass of the collision. Since then, the LHC energy has been increasing: 1.8 TeV at the end of 2009, 7 TeV for the whole of 2010 and 2011, then 8 TeV in 2012. The first data-taking period performed between 2010 and 2012 is referred to as Run I.
2. "CERN Accelerating Science," ATLAS Experiment at CERN, 2024, https://atlas.cern.
3. R. Achenbach, P. Adragna, M. Aharrouche, et al., "First Data with the Atlas Level-1 Calorimeter Trigger," CERN Document Server, November 20, 2008, https://cds.cern.ch/record/1140951?ln=en.
4. The ATLAS Level-1 Calorimeter Trigger is one of the main elements of the first stage of event selection for the ATLAS experiment at the LHC. The input stage consists of a mixed analogue/digital component taking trigger sums from the ATLAS calorimeters. The trigger logic is performed in a digital, pipelined system with several stages of processing, largely based on FPGAs, which perform programmable algorithms in parallel with a fixed latency to process about 300 GB of input data. The real-time output consists of counts of different types of physics objects, and energy sums. The final system consists of over 300 custom-built VME modules, of several different types.
5. Fabio Duarte, "Amount of Data Created Daily (2024)," *Exploding Topics*, June 13, 2024, https://explodingtopics.com/blog/data-generated-per-day.
6. Petroc Taylor, "Data Growth Worldwide 2010–2025," *Statista*, November 16, 2023, https://www.statista.com/statistics/871513/worldwide-data-created.
7. Eric Ries, *The Lean Startup* (New York: Crown Currency, 2011).
8. Jeremy Kahn, "It's About Better Data, Not Big Data, Deep Learning Pioneer Ng Says," *Fortune*, March 21, 2023, https://fortune.com/2022/06/21/andrew-ng-data-centric-ai.
9. E. L. Barse, H. Kvarnstrom, and E. Johnson, "Synthesizing Test Data for Fraud Detection Systems," *19th Annual Computer Security Applications Conference Proceedings*, December 2003, 10.1109/csac.2003.1254343.

10. Aldren Gonzales, Guruprabha Guruswamy, and Scott R. Smith, "Synthetic Data in Health Care: A Narrative Review," *PLOS Digital Health 2*, no. 1 (2023): e0000082, 10.1371/journal.pdig.0000082.
11. Dominik Ziemke, Ihab Kaddoura, and Kai Nagel, "The Matsim Open Berlin Scenario: A Multimodal Agent-Based Transport Simulation Scenario Based on Synthetic Demand Modeling and Open Data," *Procedia Computer Science* 151 (2019): 870–77, 10.1016/j.procs.2019.04.120.

Chapter 4

1. Iavor Bojinov, "Keep Your AI Projects on Track," *Harvard Business Review*, November 2023, https://hbr.org/2023/11/keep-your-ai-projects-on-track.
2. Gartner, "Gartner Poll Finds 55% of Organizations Are in Piloting or Production Mode with Generative AI," October 3, 2023, https://www.gartner.com/en/newsroom/press-releases/2023-10-03-gartner-poll-finds-55-percent-of-organizations-are-in-piloting-or-production-mode-with-generative-ai.
3. Musa Hanhan, "Why Your Predictive Analytics and AI Projects Are Failing – and How to Transform Your Success," CX Network, March 25, 2024, https://www.cxnetwork.com/artificial-intelligence/articles/why-your-predictive-analytics-and-ai-projects-are-failing-and-how-to-transform-your-success.
4. Gartner, "Gartner Says Nearly Half of CIOs Are Planning to Deploy Artificial Intelligence," February 13, 2018, https://www.gartner.com/en/newsroom/press-releases/2018-02-13-gartner-says-nearly-half-of-cios-are-planning-to-deploy-artificial-intelligence.
5. Edward Hance Shortliffe, "Design Considerations for MYCIN," in *MYCIN: A Knowledge-Based Consultation Program in Artificial Intelligence*, ed. Bruce G. Buchanan and Edward H. Shortliffe (New York: Elsevier, 1976), Chapter 2.
6. Feng-hsiung Hsu, *Behind Deep Blue: Building the Computer That Defeated the World Chess Champion* (Princeton, NJ: Princeton University Press, 2002).
7. Ivan Bratko, *PROLOG Programming for Artificial Intelligence* (Harlow, UK: Pearson Education Limited, 2011).
8. More examples include the following:
   DENDRAL study: This paper details the development of DENDRAL and its use in organic chemistry for molecular structure elucidation. DENDRAL was another early expert systems project used to learn the process of scientific hypotheses – Lindsay, Robert K., Bruce G. Buchanan, Edward A. Feigenbaum, and Joshua Lederberg. "Dendral: A Case Study

of the First Expert System for Scientific Hypothesis Formation." *Artificial Intelligence* 61, no. 2 (June 1993): 209–61. 10.1016/0004-3702(93)90068-m. SHRDLU study: This study explores SHRDLU's capabilities in understanding and processing natural language commands to manipulate a block world. SHRDLU was an early natural language understanding device developed by Terry Winograd that could respond to linguistic commands and induct from context – Winograd, Terry. *Understanding Natural Language* (New York: Academic Press, 1972).
9. Jair Cervantes, Farid Garcia-Lamont, Lisbeth Rodríguez-Mazahua, et al., "A Comprehensive Survey on Support Vector Machine Classification: Applications, Challenges and Trends," *Neurocomputing*, 415, 74–83, 10.1016/j.neucom.2019.10.118.
10. Daniel Jurafsky and James H. Martin, *Speech and Language Processing*, 2nd ed. (Upper Saddle River, NJ: Prentice Hall, 2009).
11. Debajit Datta, Preetha Evangeline David, Dhruv Mittal, et al., "Neural Machine Translation Using Recurrent Neural Network," *International Journal of Engineering and Advanced Technology* 9, no. 4 (2020): 1395–1400, 10.35940/ijeat.D7637.049420.
12. Ni, Jianjun, Yinan Chen, Yan Chen, Jinxiu Zhu, Deena Ali, and Weidong Cao. "A Survey on Theories and Applications for Self-Driving Cars Based on Deep Learning Methods," *Applied Sciences* 10, no. 8 (2020): 2749. https://doi.org/10.3390/app10082749.
13. Jacob Devlin, Mig-Wei Chang, Kenton Lee, et al., "BERT: Pre-training of Deep Bidirectional Transformers for Language Understanding," October 11, 2018, https://arxiv.org/abs/1810.04805.
14. OpenAI, "Language Unsupervised," June 11, 2018, https://openai.com/index/language-unsupervised.
15. Alex Hern, "What Is the AI Chatbot Phenomenon ChatGPT and Could It Replace Humans?" *The Guardian*, December 5, 2022, http:// https://www.theguardian.com/technology/2022/dec/05/what-is-ai-chatbot-phenomenon-chatgpt-and-could-it-replace-humans.
16. Tenten, "Unlocking the Potential of Virtual Production," February 1, 2024, https://tenten.co/insight/article/unlocking-the-potential-of-virtual-production.
17. International College of Fashion, "The Role of Generative AI in Fashion," February 22, 2024, https://www.icf.edu.in/blog/the-role-of-generative-ai-in-fashion.
18. Jenna Gibson, "AI Art at Christie's Sells for $432,500," *New York Times*, October 25, 2018, https://www.nytimes.com/2018/10/25/arts/design/ai-art-sold-christies.html.
19. Will Knight, "The Dark Secret at the Heart of AI," *MIT Technology Review*, April 11, 2017, https://www.technologyreview.com/2017/04/11/5113/the-dark-secret-at-the-heart-of-ai.

20. Gartner, "Generative AI," accessed July 2, 2024, https://www.gartner.com/en/topics/generative-ai.
21. Weiyu Ma, Qitui Mi, Xue Yan, et al., "Large Language Models Play StarCraft II: Benchmarks and A Chain of Summarization Approach," arXiv, https://arxiv.org/abs/2312.11865v1.
22. ZME Science, "How Ants Are Inspiring the Design and Optimization of Logistics Algorithms," accessed July 2, 2024, https://www.zmescience.com/science/how-ants-are-inspiring-the-design-and-optimization-of-logistics-algorithms/#:~:text=When%20they%20travel%2C%20they%20deposit,them%20more%20attractive%20to%20follow.
23. Li Zheng, Wenjie Yu, Guangxu Li, et al., "Particle Swarm Algorithm Path-Planning Method for Mobile Robots Based on Artificial Potential Fields," Sensors 2023, 23(13), 6082, 10.3390/s23136082.
24. Paul A. Merolla, John V. Arthur, Rodrigo Alvarez-Icaza, et al., "A Million Spiking-Neuron Integrated Circuit with a Scalable Communication Network and Interface," Science 345, no. 6197 (2014): 668–73, 10.1126/science.1254642.

Chapter 5

1. Craig Gillespie, director, *Dumb Money* (Sony Pictures, 2023).
2. National Retail Federation, "$428 Billion in Merchandise Returned in 2020," January 27, 2021, https://nrf.com/media-center/press-releases/428-billion-merchandise-returned-2020.
3. Accuracy measures the percentage of correctly predicted instances out of the total instances, precision measures the percentage of relevant instances among the retrieved instances, and recall measures the percentage of relevant instances that were retrieved out of all relevant instances. These metrics help in evaluating the performance and reliability of predictive models.
4. Identity resolution models use algorithms to match and consolidate customer data from various sources, eliminating duplicates and ensuring a single, unified view of each customer. This process often involves matching on attributes like name, address, phone number, email, and transaction history, and it is crucial for accurate data analysis and personalized marketing efforts.
5. A single-layer NN, also known as a perceptron, consists of an input layer directly connected to an output layer without any hidden layers in between. This simplistic structure limits the model's ability to capture complex patterns and interactions in the data.

Chapter 6

1. Daniel Kahneman, Jack L Knetsch, and Richard H Thaler, "Anomalies: The Endowment Effect, Loss Aversion, and Status Quo Bias," *Journal of Economic Perspectives* 5, no. 1 (1991): 193–206, 10.1257/jep.5.1.193.
2. Kahneman, Knetsch, and Thaler, "Anomalies" 193–206.
3. Martie G. Haselton and Daniel Nettle, "The Paranoid Optimist: An Integrative Evolutionary Model of Cognitive Biases," *Personality and Social Psychology Review* 10, no. 1 (2006): 47–66, 10.1207/s15327957pspr1001_3.
4. Daniel Kahneman, *Thinking, Fast and Slow* (London, UK: Penguin, 2012).
5. An evolutionary theory proposed by Charles Darwin, suggesting that individuals or groups best adapted to their environment are more likely to survive and reproduce. This concept is often used metaphorically to describe how well-adapted systems or practices persist over time. See Charles Darwin, Paul H. Barrett, and R. B. Freeman, *On the Origin of Species, 1859* (London: Routledge, 2016).
6. Colin Camerer, George Loewenstein, and Martin Weber, "The Curse of Knowledge in Economic Settings: An Experimental Analysis," *Journal of Political Economy* 97, no. 5 (1989): 1232–54, 10.1086/261651.
7. Peer effects refer to the influence that peers have on each other's attitudes and behaviors, often leading individuals to conform to the group to avoid being left out or judged. This social influence can drive individuals to align their actions with the perceived norms of their peer group.
8. A commitment device is a technique used to bind oneself to a plan of action that one might not follow through on due to lack of willpower or changing preferences, thus increasing the likelihood of achieving a goal. It helps ensure that intentions are turned into actions by creating a form of self-imposed constraint or incentive.

Additional Reading

Additional Reading on Applications of Classical Machine Learning

DENDRAL study: This paper details the development of DENDRAL and its use in organic chemistry for molecular structure elucidation. DENDRAL was another early expert systems project used to learn the process of scientific hypotheses – Lindsay, Robert K., Bruce G. Buchanan, Edward A. Feigenbaum, and Joshua Lederberg. "Dendral: A Case Study of the First Expert System for Scientific Hypothesis Formation." *Artificial Intelligence* 61, no. 2 (June 1993): 209–61. 10.1016/0004-3702(93)90068-m.

SHRDLU study: This study explores SHRDLU's capabilities in understanding and processing natural language commands to manipulate a block world. SHRDLU was an early natural language understanding device developed by Terry Winograd that could respond to linguistic commands and induct from context – Winograd, Terry. *Understanding Natural Language* (New York: Academic Press, 1972).

Additional Reading on Applications of Machine Learning

Natural language processing (NLP) study: This study evaluated various feature engineering techniques for classifying the polarity of patient-authored texts in online health forums. It demonstrates that word embeddings significantly outperform traditional bag-of-words representations. Additionally, the study highlights the importance of "polar facts" (objective information with positive or negative connotations) in sentiment analysis within the health domain. Carrillo-de-Albornoz, Jorge, Javier Rodríguez Vidal, and Laura Plaza. "Feature Engineering for Sentiment Analysis in E-Health Forums." *PLOS ONE* 13, no. 11 (November 29, 2018). 10.1371/journal.pone.0207996.

Feature Modelling study: This deep-learning-based model combines extensive feature engineering and a customized long short-term memory neural network to analyze data from the Chinese stock market. The proposed system includes data preprocessing and multiple feature engineering techniques, leading to high accuracy in short-term price trend predictions. Shen, Jingyi, and M. Omair Shafiq. "Short-Term Stock Market Price Trend Prediction Using a Comprehensive Deep Learning System." *Journal of Big Data* 7, no. 1 (August 28, 2020). 10.1186/s40537-020-00333-6.

Image Recognition study: This study discusses a method combining convolutional neural networks (CNNs), principal component analysis (PCA), and support vector classification (SVC) to enhance face recognition accuracy. Initially, CNNs are employed to extract discriminative feature vectors from face images. These features are then reduced in dimensionality using PCA, which helps in retaining essential information while minimizing computational load. SVC is used for classification, leading to improved recognition performance. This integrated approach leverages the strengths of each technique to achieve robust face recognition results. Liu, Yu Han. "Feature Extraction and Image Recognition with Convolutional Neural Networks." *Journal of Physics: Conference Series* 1087 (September 2018): 062032. 10.1088/1742-6596/1087/6/062032.

Health care study: In this study, researchers Elkin and Zhu use machine learning to predict the termination of clinical trials. By analyzing 311 260 trials and creating 640 features, they identify key factors such as trial eligibility and sponsor types related to terminations. Their model achieves over 67% balanced accuracy and a 0.73 AUC (area under the curve) score, demonstrating the effectiveness of machine learning in forecasting clinical trial outcomes. Elkin, Magdalyn E., and Xingquan Zhu. "Predictive Modeling of Clinical Trial Terminations Using Feature Engineering and Embedding Learning." *Scientific Reports* 11, no. 1 (February 10, 2021). 10.1038/s41598-021-82840-x.

Additional Reading on Deep Learning Models

Image recognition study: This study introduces AlexNet, a deep convolutional neural network that significantly advanced the state of the art in image classification. By leveraging graphics processing units for training, rectified linear unit activation functions, and dropout for regularization, AlexNet achieved a top-five error rate of 15.3% on the ImageNet dataset, a notable improvement over previous methods. Krizhevsky, Alex, Ilya Sutskever, and Geoffrey E. Hinton. "ImageNet Classification with Deep Convolutional Neural Networks." *Communications of the ACM* 60, no. 6 (2017): 84–90.

NLP study: Vaswani et al. introduce the Transformer model, a novel neural network architecture based solely on attention mechanisms, discarding recurrence and convolutions entirely. The Transformer achieves superior performance in tasks such as machine translation, exemplified by its state-of-the-art results on the WMT 2014 English-to-German and English-to-French translation benchmarks. This architecture's key innovation is the self-attention mechanism, which enables models to process entire input sequences simultaneously, significantly improving training efficiency and performance. Vaswani, Ashish, Noam Shazeer, Miki Parmar, et al. "Attention Is All You Need." *Advances in Neural Information Processing Systems* 30 (2017): 5998–6008.

Speech recognition study: This paper presents a highly accurate speech recognition system using deep learning. It leverages large recurrent neural networks trained on extensive datasets, achieving superior performance (especially in noisy environments) compared to traditional models. Hannun, Awni, et al. "Deep Speech: Scaling Up End-to-End Speech Recognition." *arXiv preprint arXiv:1412.5567* (2014).

Autonomous driving study: The study by Bojarski et al. describes a CNN trained to map raw pixels from a single front-facing camera directly to steering commands. The end-to-end approach enables the system to drive on many different roads and in many conditions by learning necessary internal representations from the data without explicit programming. Bojarski, Mariusz, Davide Del Testa, Daniel Dworakowski, et al. "End to End Learning for Self-Driving Cars." *arXiv preprint arXiv:1604.07316* (2016).

Random forests study: This study introduces risk-controlled decision trees and risk-controlled random forests, which provide interpretable models for generating individualized treatment rules. By integrating risk constraints into the decision-making process, these models offer robust solutions for precision medicine. Doubleday, Kevin, Jin Zhou, Hua Zhou, and Haoda Fu. "Risk Controlled Decision Trees and Random Forests for Precision Medicine." *Statistics in Medicine* 41, no. 4 (November 16, 2021): 719–35. 10.1002/sim.9253.

ResNet image recognition study: This study introduces ResNet, an architecture that uses shortcut connections to address the vanishing gradient problem, allowing for the training of very deep networks. ResNet won the 2015 ImageNet Large Scale Visual Recognition Challenge. He, Kaiming, Xiangyu Zhang, Shaoqing Ren, et al. "Deep Residual Learning for Image Recognition." *Proceedings of the IEEE Conference on Computer Vision and Pattern Recognition* (CVPR) (2016): 770–78.

NLP machine and deep learning comparison: This study evaluates text classification methods and reveals that deep learning techniques, particularly CNNs and RNNs, outperform traditional methods such as SVM and Naive Bayes in both accuracy and scalability. Kamath, Cannannore Nidhi, Syed Saqib Bukhari, and Andreas Dengel. "Comparative Study Between Traditional Machine Learning and Deep Learning Approaches for Text Classification." *Proceedings of the ACM Symposium on Document Engineering 2018*, August 28, 2018. 10.1145/3209280.3209526.

Additional Reading on Generative AI

Generative adversarial nets (GANs) for image generation: This paper introduces GANs, where two neural networks – a generator and a discriminator – compete to create and evaluate synthetic data, respectively. This approach allows for the generation of highly realistic synthetic data, revolutionizing applications in image synthesis and data augmentation. Goodfellow,

Ian, Jean Pouget-Abadie, Mehdi Mirza, et al. "Generative Adversarial Nets." *Advances in Neural Information Processing Systems* 27 (2014): 2672–80.

Transformers for text generation: This study trained GPT-3, an autoregressive language model with 175 billion parameters. GPT-3 demonstrates strong few-shot learning capabilities across various NLP tasks without task-specific fine-tuning, performing competitively with state-of-the-art models. The model excels in tasks such as translation, question answering, and text generation, showcasing the potential of large-scale language models under minimal supervision. Brown, Tom B., Benjamin Mann, Nick Ryder, et al. "Language Models are Few-Shot Learners." *Advances in Neural Information Processing Systems* 33 (2020): 1877–1901.

Deepfakes and video synthesis study: This study introduces BERT, a transformer-based model pretrained on a large corpus using masked language modeling and next sentence prediction tasks. BERT significantly improves performance on various NLP benchmarks, including *GLUE*, SQuAD, and SWAG, by providing deep bidirectional representations. This model can be fine-tuned for specific tasks with minimal adjustments, demonstrating its versatility and effectiveness in understanding language context. Kim, Hyeongwoo, Pablo Garrido, Ayush Tewari, et al. "Deep Video Portraits." *ACM Transactions on Graphics* 37, no. 4 (2018): 163.

GANS for customer data: This study explores how GANs, used for generating fashion designs, influence consumer perceptions and purchasing intentions. The findings reveal that GAN-generated fashion items can positively affect consumers' attitudes toward innovation and creativity, leading to higher engagement and purchase likelihood. Sohn, Kwonsang, Christine Eunyoung Sung, Gukwon Koo, and

Ohbyung Kwon. "Artificial Intelligence in the Fashion Industry: Consumer Responses to Generative Adversarial Network (GAN) Technology." *International Journal of Retail & Distribution Management* 49, no. 1 (September 21, 2020): 61–80. 10.1108/ijrdm-03-2020-0091.

Synthetic data for market analysis: The study evaluates various methods for generating and assessing synthetic data, emphasizing GANs' ability to mimic complex data distributions. It highlights GANs' applications in areas such as data augmentation for machine learning, privacy-preserving data sharing, and enhancing training datasets for better model performance. Figueira, Alvaro, and Bruno Vaz. "Survey on Synthetic Data Generation, Evaluation Methods and Gans." *Mathematics* 10, no. 15 (August 2, 2022): 2733. 10.3390/math10152733.

Additional Reading on Deep Learning and Biologically Inspired Models

Convolutional neural network study: This work presents a large-scale image classification model using deep convolutional neural networks. The method uses a network architecture composed of five convolutional layers and three fully connected layers, optimized with nonsaturating neurons and a GPU-accelerated implementation. This approach enabled the classification of 1.2 million high-resolution images from the ImageNet dataset into 1000 classes. Through the use of dropout regularization to mitigate overfitting, the model achieved top-tier performance with error rates significantly better than the previous state-of-the-art, setting a new benchmark. Krizhevsky, Alex, Ilya Sutskever, and Geoffrey E. Hinton. "ImageNet Classification with Deep Convolutional Neural Networks." *Communications of the ACM* 60, no. 6 (2017): 84–90.

Reinforcement learning study: This paper presents a biologically inspired approach using deep Q-networks. This method combines deep learning with reinforcement learning, enabling an artificial agent to achieve human-level performance in Atari games by mimicking human learning processes through reward-based decision-making. Mnih, Volodymyr, Koray, Kavukcuoglu, David Silver, et al. "Human-Level Control Through Deep Reinforcement Learning." *Nature* 518, no. 7540 (2015): 529–33.

NeuroEvolution of augmenting topologies: This paper introduces the NEAT (NeuroEvolution of Augmenting Topologies) algorithm, which evolves neural networks by optimizing both their weights and structures. This biologically inspired approach starts with simple networks and gradually complexifies them, demonstrating improved performance in various tasks by evolving more sophisticated topologies over generations. Stanley, Kenneth O., and Risto Miikkulainen. "Evolving Neural Networks Through Augmenting Topologies." *Evolutionary Computation* 10, no. 2 (2002): 99–127.

Spiking neural networks (SNNs): This paper discusses the advantages of SNNs in energy efficiency and real-time processing, highlighting their applications in robotics, sensory processing, and neuromorphic computing. The review also addresses current challenges and future research directions for advancing SNN technology. Yamazaki, Kashu, Viet-Khoa Vo-Ho, Darshan Bulsara, and Ngan Le. "Spiking Neural Networks and Their Applications: A Review." *Brain Sciences* 12, no. 7 (2022): 863. 10.3390/brainsci12070863.

Bibliography

Achenbach, R., P. Adragna, M. Aharrouche, et al. "First Data with the Atlas Level-1 Calorimeter Trigger," CERN Document Server, November 20, 2008, https://cds.cern.ch/record/1140951?ln=en.

Ariely, Dan. "Combining Experiences over Time: The Effects of Duration, Intensity Changes and On-Line Measurements on Retrospective Pain Evaluations," *Journal of Behavioral Decision Making* 11, no. 1 (1998): 19–45, 10.1002/(sici)1099-0771(199803)11:1<19::aid-bdm277>3.0.co;2-b.

Ariely, Dan. *Predictably Irrational: The Hidden Forces That Shape Our Decisions* (London: Harper Collins, 2008).

Ariely, Dan. "Will Pay for Performance Backfire? Insights from Behavioral Economics," *Forefront Group*, October 11, 2012, 10.1377/forefront.20121011.023909.

Ariely, Dan. "Column: Why Businesses Don't Experiment," *Harvard Business Review*, August 1, 2014, https://hbr.org/2010/04/column-why-businesses-dont-experiment.

Ariely, Dan, and Klaus Wertenbroch. "Procrastination, Deadlines, and Performance: Self-Control by Precommitment," *Psychological Science* 13, no. 3 (2002): 219–24, 10.1111/1467-9280.00441.

Bahcall, Safi. *Loonshots: How to Nurture the Crazy Ideas That Win Wars, Cure Diseases, and Transform Industries* (New York: St. Martin's Press, 2019).

Barse, E. L., H. Kvarnstrom, and E. Johnson. "Synthesizing Test Data for Fraud Detection Systems," *19th Annual Computer Security Applications Conference Proceedings*, December 2003, 10.1109/csac.2003.1254343.

Baymard Institute, "Average Cart Abandonment Rate," April 2023, https://baymard.com/lists/cart-abandonment-rate.

Blaga, Oana M., Livia Vasilescu, and Razvan M. Chereches. "Use and Effectiveness of Behavioural Economics in Interventions for Lifestyle Risk Factors of Non-Communicable Diseases: A Systematic Review with

Policy Implications," *Perspectives in Public Health* 138, no. 2 (2017): 100–10, 10.1177/1757913917720233.

Bojinov, Iavor. "Keep Your AI Projects on Track," *Harvard Business Review*, November 2023, https://hbr.org/2023/11/keep-your-ai-projects-on-track.

Bratko, Ivan. *PROLOG Programming for Artificial Intelligence*, 4th ed. (Harlow, UK: Pearson Education Limited, 2011).

Camerer, Colin, George Loewenstein, and Martin Weber. "The Curse of Knowledge in Economic Settings: An Experimental Analysis," *Journal of Political Economy* 97, no. 5 (1989): 1232–54, 10.1086/261651.

"CERN Accelerating Science," ATLAS Experiment at CERN, 2024, https://atlas.cern/.

Cervantes, Jair, Farid Garcia-Lamont, Lisbeth Rodríguez-Mazahua, et al. "A Comprehensive Survey on Support Vector Machine Classification: Applications, Challenges and Trends," *Neurocomputing*, 415, 74–83, 10.1016/j.neucom.2019.10.118.

Chamberlin, Edward H. "An Experimental Imperfect Market," *Journal of Political Economy* 56, no. 2 (1948): 95–108, 10.1086/256654.

Chang, Leslie L., Adam D. DeVore, Bradi B. Grander, et al. "Leveraging Behavioral Economics to Improve Heart Failure Care and Outcomes," *Circulation* 136, no. 8 (2017): 765–72, 10.1161/circulationaha.117.028380.

Christensen, Clayton. *The Innovator's Dilemma: The Revolutionary Book That Will Change the Way You Do Business* (New York: HarperCollins, 2003).

Darwin, Charles, Paul H. Barrett, and R. B. Freeman. *On the Origin of Species, 1859* (London: Routledge, 2016).

Datta, Debajit, Preetha Evangeline David, Dhruv Mittal, et al. "Neural Machine Translation Using Recurrent Neural Network," *International Journal of Engineering and Advanced Technology* 9, no. 4 (2020): 1395-1400, 10.35940/ijeat.D7637.049420.

Dennis, Alan R., Lingyao Yuan, Xuan Feng, et al. "Digital Nudging: Numeric and Semantic Priming in e-Commerce," *Journal of Management Information Systems* 37, no. 1 (2020): 39–65, 10.1080/07421222.2019.1705505.

Devlin, Jacob, Mig-Wei Chang, Kenton Lee, et al. "BERT: Pre-Training of Deep Bidirectional Transformers for Language Understanding," October 11, 2018, https://arxiv.org/abs/1810.04805.

Duarte, Fabio. "Amount of Data Created Daily (2024)," *Exploding Topics*, June 13, 2024, https://explodingtopics.com/blog/data-generated-per-day.

Gartner, "Gartner Says Nearly Half of CIOs Are Planning to Deploy Artificial Intelligence," February 13, 2018, https://www.gartner.com/en/newsroom/press-releases/2018-02-13-gartner-says-nearly-half-of-cios-are-planning-to-deploy-artificial-intelligence.

Bibliography

Gartner, "Gartner Poll Finds 55% of Organizations Are in Piloting or Production Mode with Generative AI," October 3, 2023, https://www.gartner.com/en/newsroom/press-releases/2023-10-03-gartner-poll-finds-55-percent-of-organizations-are-in-piloting-or-production-mode-with-generative-ai.

Gartner, "Generative AI," accessed July 2, 2024, https://www.gartner.com/en/topics/generative-ai.

Gibson, Jenna. "AI Art at Christie's Sells for $432,500," *New York Times*, October 25, 2018, https://www.nytimes.com/2018/10/25/arts/design/ai-art-sold-christies.html.

Gillespie, Craig, director. *Dumb Money* (Sony Pictures, 2023).

Gino, Francesca. "Uber Shows How Not to Apply Behavioral Economics," *Harvard Business Review*, April 13, 2017, https://hbr.org/2017/04/uber-shows-how-not-to-apply-behavioral-economics.

Gonzales, Aldren, Guruprabha Guruswamy, and Scott R. Smith, "Synthetic Data in Health Care: A Narrative Review," *PLOS Digital Health* 2, no. 1 (2023): e0000082, 10.1371/journal.pdig.0000082.

Guinart, Daniel. and John M. Kane, "Use of Behavioral Economics to Improve Medication Adherence in severe Mental Illness," *Psychiatric Services* 70, no. 10 (2019): 955–57, 10.1176/appi.ps.201900116.

Hanhan, Musa. "Why Your Predictive Analytics and AI Projects Are Failing – and How to Transform Your Success," CX Network, March 25, 2024, https://www.cxnetwork.com/artificial-intelligence/articles/why-your-predictive-analytics-and-ai-projects-are-failing-and-how-to-transform-your-success.

Haselton Martie G., and Daniel Nettle. "The Paranoid Optimist: An Integrative Evolutionary Model of Cognitive Biases," *Personality and Social Psychology Review* 10, no. 1 (2006): 47–66, 10.1207/s15327957pspr1001_3.

Hern, Alex. "What Is the AI Chatbot Phenomenon ChatGPT and Could It Replace Humans?" *The Guardian*, December 5, 2022, https://www.theguardian.com/technology/2022/dec/05/what-is-ai-chatbot-phenomenon-chatgpt-and-could-it-replace-humans.

Herrett, Emily, Tjeerd van Staa, Caroline Free, et al. "Text Messaging Reminders for Influenza Vaccine in Primary Care: Protocol for a Cluster Randomised Controlled Trial (TXT4FLUJAB)," *BMJ Open* 4, no. 5 (2014), 10.1136/bmjopen-2013-004633.

Heyman, James, and Dan Ariely. "Effort for Payment," *Psychological Science* 15, no. 11 (2004): 787–93, 10.1111/j.0956-7976.2004.00757.x.

Horne, Benjamin D., Joseph B. Muhlestein, Donald L. Lappé, et al. "Behavioral Nudges as Patient Decision Support for Medication Adherence: The

Encourage Randomized Controlled Trial," *American Heart Journal* 244 (2022): 125–34, 10.1016/j.ahj.2021.11.001.

Hsu, Feng-hsiung. *Behind Deep Blue: Building the Computer that Defeated the World Chess Champion*, (Princeton, NJ: Princeton University Press, 2002).

International College of Fashion. "The Role of Generative AI in Fashion," February 22, 2024, https://www.icf.edu.in/blog/the-role-of-generative-ai-in-fashion/.

Jackson, Matthew. *The Social and Economic Networks* (Princeton, NJ: Princeton University Press, 2008).

Jahoda, Gustav. "A Cross-Cultural Perspective on Developmental Psychology," *International Journal of Behavioral Development* 9, no. 4 (1986): 417–37, 10.1177/016502548600900402.

Juneja, Prerna, and Tanushree Mitra. "Auditing E-Commerce Platforms for Algorithmically Curated Vaccine Misinformation," *Proceedings of the 2021 CHI Conference on Human Factors in Computing Systems*, May 6, 2021, 10.1145/3411764.3445250.

Jurafsky, Daniel, and James H. Martin. *Speech and Language Processing*, 2nd ed. (Upper Saddle River, NJ: Prentice Hall, 2009).

Kahn, Jeremy. "It's About Better Data, Not Big Data, Deep Learning Pioneer Ng Says," *Fortune*, March 21, 2023, https://fortune.com/2022/06/21/andrew-ng-data-centric-ai/.

Kahneman, Daniel. *Thinking Fast and Slow* (New York: Penguin Books, 2011).

Kahneman, Daniel, Jack L. Knetsch, and Richard H. Thaler. "Anomalies: The Endowment Effect, Loss Aversion, and Status Quo Bias," *Journal of Economic Perspectives* 5, no. 1 (1991): 193–206, 10.1257/jep.5.1.193.

Kahneman, Daniel, Olivier Sibony, and Cass R. Sunstein. *Noise: A Flaw in Human Judgment* (New York: Little, Brown Spark, 2022).

Kahneman Daniel, and Amos Tversky. "Prospect Theory: An Analysis of Decision Under Risk," *Econometrica* 47, no. 2 (1979): 263, 10.2307/1914185.

Kala, Devkant, Dhani Shanker Chaubey, and Ahmad Samed Al-Adwan. "Cryptocurrency Investment Behaviour of Young Indians: Mediating Role of Fear of Missing Out," *Global Knowledge, Memory and Communication*, 2023, 10.1108/gkmc-07-2023-0237.

Katona, George. *Psychological Analysis of Economic Behaviour* (Westport, CT: Greenwood Press, 1977).

Keynes, John Maynard. *The General Theory of Employment, Interest and Money* (London: Palgrave Macmillan, 1936).

Knight, Will. "The Dark Secret at the Heart of AI," *MIT Technology Review*, April 11, 2017, https://www.technologyreview.com/2017/04/11/5113/the-dark-secret-at-the-heart-of-ai/.

Bibliography

Kolb, Robert. "Pareto, Vilfredo (1848–1923)," *Encyclopedia of Business Ethics and Society* (Thousand Oaks, CA: SAGE Publications, 2008).

Lee, Chang-Yuan, Carey K. Morewedge, Guy Hochman, et al. "Small Probabilistic Discounts Stimulate Spending: Pain of Paying in Price Promotions," *Journal of the Association for Consumer Research* 4, no. 2 (2019): 160–71, 10.1086/701901.

Lindsay, Robert K., Bruce G. Buchanan, Edward A. Feigenbaum, and Joshua Lederberg. "Dendral: A Case Study of the First Expert System for Scientific Hypothesis Formation." *Artificial Intelligence* 61, no. 2 (1993): 209–61, 10.1016/0004-3702(93)90068-m.

Loxton, Mary, Robert Truskett, Brigitte Scarf, et al. "Consumer Behaviour During Crises: Preliminary Research on How Coronavirus Has Manifested Consumer Panic Buying, Herd Mentality, Changing Discretionary Spending and the Role of the Media in Influencing Behaviour," *Journal of Risk and Financial Management* 13, no. 8 (2020): 166, https://www.mdpi.com/1911-8074/13/8/166.

Ma, Weiyu, Qitui Mi, Xue Yan, et al., "Large Language Models Play Starcraft II: Benchmarks and a Chain of Summarization Approach," arXiv, https://arxiv.org/abs/2312.11865v1.

McAfee, Andrew. *The Geek Way: The Radical Mindset That Drives Extraordinary Results* (New York: Little, Brown, 2023).

Merolla, Paul A., John V. Arthur, Rodrigo Alvarez-Icaza, et al. "A Million Spiking-Neuron Integrated Circuit with a Scalable Communication Network and Interface," *Science* 345, no. 6197 (2014): 668-73, 10.1126/science.1254642.

National Retail Federation. "$428 Billion in Merchandise Returned in 2020," January 27, 2021, https://nrf.com/media-center/press-releases/428-billion-merchandise-returned-2020.

Ni, Jianjun, Yinan Chen, Yan Chen, Jinxiu Zhu, Deena Ali, and Weidong Cao. "A Survey on Theories and Applications for Self-Driving Cars Based on Deep Learning Methods." *Applied Sciences* 10, no. 8 (2020): 2749. https://doi.org/10.3390/app10082749.

OpenAI, "Language Unsupervised," June 11, 2018, https://openai.com/index/language-unsupervised/.

Peterson, Richard L. "The Neuroscience of Investing: fMRI of the Reward System," *Brain Research Bulletin* 67, no. 5 (2005): 391–97, 10.1016/j.brainresbull.2005.06.015.

Ries, Eric. *The Lean Startup* (New York: Crown Currency, 2011).

Rutter, Jill. "'Nudge Unit,'" Institute for Government, March 2, 2010, https://www.instituteforgovernment.org.uk/article/explainer/nudge-unit.

Scheiber, Noam. "How Uber Uses Psychological Tricks to Push Its Drivers' Buttons," *New York Times*, April 2, 2017, https://www.nytimes.com/interactive/2017/04/02/technology/uber-drivers-psychological-tricks.html.

Sharma, Arpana, Pruthi Madhu, and Sageena Geetanjali. "Adoption of Telehealth Technologies: An Approach to Improving Healthcare System," *Translational Medicine Communications* 7, no. 1 (2022): 20, 10.1186/s41231-022-00125-5.

Shortliffe, Edward Hance. "Design Considerations for MYCIN," in *MYCIN: A Knowledge-Based Consultation Program in Artificial Intelligence*, ed. Bruce G. Buchanan and Edward H. Shortliffe (New York: Elsevier, 1976), pp. 20–54.

Simon, Herbert A., and Robert Dahl. "Administrative Behavior: A Study of Decision-Making Processes in Administrative Organization," *Administrative Science Quarterly* 2, no. 2 (1957): 244–48, 10.2307/2390693.

Sloan, P. A., and Irving Fisher. "The Theory of Interest," *The Economic Journal* 41, no. 161 (1931): 84, 10.2307/2224140.

Smironva, Ekaterina, Kiattipoom Kiatkawsin, Seul Ki Lee, et al. "Self-Selection and Non-Response Biases in Customers' Hotel Ratings – A Comparison of Online and Offline Ratings," *Current Issues in Tourism* 23, no. 10 (2019): 1191–1204, 10.1080/13683500.2019.1599828.

Smith, Vernon L. *Economics in the Laboratory* (Tucson, AZ: American Economic Association, 1994).

Taylor, Petroc. "Data Growth Worldwide 2010-2025," *Statista*, November 16, 2023, https://www.statista.com/statistics/871513/worldwide-data-created/.

Tenten. "Unlocking the Potential of Virtual Production," February 1, 2024, https://tenten.co/insight/article/unlocking-the-potential-of-virtual-production/.

Thaler, Richard H. "The End of Behavioral Finance," *Financial Analysts Journal* 55, no. 6 (1999): 12–17, 10.2469/faj.v55.n6.2310.

Thaler, Richard H. "From Homo Economicus to Homo Sapiens," *Journal of Economic Perspectives* 14, no. 1 (2000): 133–41, 10.1257/jep.14.1.133.

Thaler, Richard H. "Mental Accounting and Consumer Choice," *Marketing Science* 27, no. 1 (2008): 15–25, 10.1287/mksc.1070.0330.

Thaler, Richard H., and Shlomo Benartzi. "Save More Tomorrow™: Using Behavioral Economics to Increase Employee Saving," *Journal of Political Economy* 112, no. S1 (2004). 10.1086/380085.

Thaler, Richard H., and Cass Sunstein. *Nudge: Improving Decisions about Health, Wealth, and Happiness* (New York: Penguin Books, 2009).

Thaler, Richard H., Cass R. Sunstein, and John P. Balz. "Choice Architecture," *The Behavioral Foundations of Public Policy*, ed. Eldar Shafir (Princeton, NJ: Princeton University Press, 2013), 428–39.

Bibliography

Tversky, Amos, and Daniel Kahneman, "Judgment Under Uncertainty: Heuristics and Biases," *Judgment Under Uncertainty*, ed. Daniel Kahneman, Paul Slovic, and Amos Tversky (Cambridge, UK: Cambridge University Press, 1982), 3–20.

Veblen, Thorstein, and Martha Banta. *The Theory of the Leisure Class* (Oxford, UK: Oxford University Press, 2009).

Viswanadham, Ratnalekha V. N. "How Behavioral Economics Can Inform the Next Mass Vaccination Campaign: A Narrative Review," *Preventive Medicine Reports* 32 (2023): 102118, 10.1016/j.pmedr.2023.102118.

Winograd, Terry. *Understanding Natural Language* (New York: Academic Press, 1972).

Zheng, Li, Wenjie Yu, Guangxu Li, et al. "Particle swarm Algorithm Path-Planning Method for Mobile Robots Based on Artificial Potential Fields," *Sensors* 23, no. 13 (2023): 6082, 10.3390/s23136082.

Ziemke, Dominik, Ihab Kaddoura, and Kai Nagel. "The Matsim Open Berlin Scenario: A Multimodal Agent-Based Transport Simulation Scenario Based on Synthetic Demand Modeling and Open Data," *Procedia Computer Science* 151 (2019): 870–77, 10.1016/j.procs.2019.04.120.

ZME Science. "How Ants Are Inspiring the Design and Optimization of Logistics Algorithms," accessed July 2, 2024, https://www.zmescience.com/science/how-ants-are-inspiring-the-design-and-optimization-of-logistics-algorithms/#:~:text=When%20they%20travel%2C%20they%20deposit,them%20more%20attractive%20to%20follow.

Acknowledgments

As an immigrant, student, and entrepreneur navigating the diverse worlds of academia, nongovernmental organizations, and business, I've been incredibly fortunate to have the support of so many wonderful people who helped me, championed me, and cheered me on.

Maria Cutchey, Jim and Shelley Prier, Victor Chan, and Susanne Martin, my guardian angels, you became the family I left behind. Your unwavering love and support have been my anchor, and for that I am eternally grateful.

To my dear friends and advisors, Anke Kessler, Alex Karakanov, Arthur Robson, and Matthew Jackson. Thank you for your wisdom and patience and enduring my endless questions!

To Andrew Mack, I learned the art of asking the right questions from you. Thank you for telling me to leave academia!

To my remarkable advisors and partners, Don Berman, Nitin Rakesh, Tony Bradshaw, Meng Chee, Lokesh Chaudhry, Sylvia Fouhy, Alex Goblin, David Flaks, Nelson Sandhu, Ray Fager, Filipe Nogueira, Gary Bogdani, Samir Gami, Emily Chu, and Mike Henry, Britt Ewen, Albert Shum, Amir Banifatemi – your steadfast support and openness to my wild ideas have been the foundation of my journey. Your belief in me, your cheers, and your mantra, "just be yourself," have been my constant encouragement and the light in my path. Thank you for seeing me and for always being in my corner.

I feel incredibly lucky to work with the most intellectually diverse and curious team, always striving for excellence and pushing the boundaries of what's possible. Ted Violini, my cofounder and COO, you have been there since the beginning, always had my back, and believed in me. Miles Nurse, Fatemeh Renani, Julian Vahl, Navjot Kaur, Nerea Ruiz, Navid Siami, Will Zich, Marzi Emami, Hamed Dastranj, Tyler Schwartz, Damiano Truzzi, Cemre Kokowski, Milad Khatib Shahidi, Fateme Najafi, Hardik Sahi, Tara Nisha, Himanshu Sharma, Tao Samis-Violini, Kiera Lukomskyj, Aramie Ewen and all the team members, past and present – you are my constant source of inspiration. For you, difficult takes a day, impossible takes a week!

Finally, to my parents and brother, your love and encouragement have spanned oceans and continents. You have always been my greatest cheerleaders, and for that, my heart is forever full.

About the Author

Dr. Rogayeh Tabrizi is a distinguished leader in the field of artificial intelligence and data analytics, recognized as the AI Leader of the Year 2023 by WLDA. She is the CEO and founder of Theory+Practice, where she combines her expertise in experimental particle physics and economics to deliver innovative solutions and drive positive societal change. Dr. Tabrizi earned her PhD in Economics from Simon Fraser University, where she also studied experimental particle physics during her MSc in physics and contributed to the ATLAS Detector project at CERN.

Her interdisciplinary approach bridges the analytical rigor of physics with the strategic insights of behavioral economics, focusing on identifying and leveraging data trends to meet organizational goals. This synergy of skills led her to establish Theory+Practice, a firm dedicated to helping businesses harness the power of their data.

Dr. Tabrizi is passionate about advancing women's leadership and championing equality worldwide. She serves on the board of the International Women's Forum Canada and is actively involved in several committees dedicated to these causes. She is also a director of the Artificial Intelligence network of British Columbia (AInBC), working to position BC as a global leader in AI and ML.

Dr. Tabrizi's dedication to social good extends to her extensive work in the nonprofit sector, including leadership positions with Engineers Without Borders, Human Security Research Project, and the Dalai Lama Center for Peace and Education. Her numerous honors include the RBC Canadian Women Entrepreneur Awards – Ones to Watch Award (2020), Forty under 40 Award 2022, RBC Women of Influence Startup Finalist (2022), recognition as one of the Top 100 Data and Analytic Leaders (2022), and inclusion in the Top 50 Artificial Intelligence CEOs of 2021.

A resident of Vancouver, Canada, Dr. Tabrizi is also a world traveler with a deep appreciation for traditional Persian carpets. She speaks Turkish, English, and Persian fluently, and is committed to fostering social good through her professional and personal endeavors.

Index

Page numbers followed by *f* refer to figures.

A

A/B testing, 16, 170
Actionable insights, 3
Active learning, prioritizing, 162
Adaptive experimentation, 17
Advertisements, on Instagram, 10
Africa, xiv
Agent-based synthetic data, 65
Alexa, 82, 84
Algebra, xiii
Algebraic topology, xiii
Amazon, 16–17, 84, 111
Amazon Web Services, 157
Anchoring effect, 13, 27, 45
Anchoring heuristic, 36
"Animal spirits," 21
Applications of AI and ML, 95–135
    customer journeys, mapping, 104–111
    and customer patterns, 101–104
    forecasting, 127–135
    personalized pricing, 121–127
    propensity models, 118–121
    recommendation models, 111–118
    and segmentation, 97–104
    and small-scale experimentation, 97

Ariely, Dan, 26–30, 50–51
Art (artworks), 88
Artificial intelligence (AI), 1–3, 18, 73–76. *See also* Machine learning (ML)
    classical AI, 76–78
    data-centric approach to, 58
    generative AI, 86–90
    overview of, 75*f*
    timeline of key milestones in, 91*f*
Assumptions, hidden, 4, 5
ATLAS detector, xiii, 54–56
Auto insurance, 31
Automotive industry, 84–85
Availability heuristic, 22, 36, 39, 45

B

Bahcall, Safi, xiv
Banks, individuals' use of multiple, 30–32
Bass model, 70
Bayesian optimization, 132
Bayes model, 81–82
Behavioral economics, 3, 5, 14–18. *See also* Decision-making
    cross-cultural, 28
    current research in, 26–30
    experimentation in, 46–51
    and financial decision-making, 28–30
    history and origins of, 20–23

Behavioral economics (*continued*)
  integration of, into mainstream economics, 23–26
  in personalized pricing strategies, 124–127
  traditional vs., 20
Behavioral psychology, 92
Behavioral segmentation, *see* Segmentation
Benartzi, Shlomo, 25, 43
Bias(es), 12, 20, 22, 32, 33, 35–39, 89
Bidirectional encoder representations from transformers (BERT), 85
Big data, 1, 57–58, 78
Biologically inspired models, 90, 92–94
"Black boxes," 11, 13
Brain (brain activity), 29, 92, 93, 115
Branson, Richard, 155
Brazil, 147

C
Camerer, Colin, 144
Cameron, David, 24
Canada, 108, 145, 152
Carrying costs, 9
Cash payments, 29
Category killers, 114–116
CERN, xiii, xiv, xvii, xviii, 53, 55–56
Chamberlin, Edward, 46
Change, resistance to, 140–145
Chatbots, 85, 87
ChatGPT, 87, 89
Checking accounts, 10, 12
Choice architecture, 42, 45
Choice overload, 27
Christie's, 88
Classical AI, 76–78. *See also* Artificial intelligence (AI)
Classical economics, 20
Clothing design, 88
Cognitive shortcuts, *see* Heuristics

Collaborative filtering, 113
Commitment devices, 28
Communication, effective, 159–161
Competition, perfect, 46
Complexity, simplifying, 6–8
Confounding variables, 14
Conspicuous consumption, 21
Consulting firms, 3
Consumer behavior, decoding, 11–14
Consumer packaged goods (CPG) industry, 6, 9–10, 128, 131, 167
Content creation, 88–89
Contrast effect, 44–45
Convolutional neural networks (CNNs), 83–84
Cook, Scott, 51
Copyright, 89
COVID-19 pandemic, 1–2, 6, 15–16, 25, 69, 106, 127–128
Credit cards, 31
Credit ratings, 122–123
Credit risk, assessing, 79
Cross-cultural behavioral economics, 28
Cuban, Mark, 73
Cultural norms, 12
Curiosity, 161–163
"Curse of knowledge," 144, 150
Customers:
  asking, to open an account, 4
  changing preferences/behaviors of, 15
  focus on, 5
  needs and preferences of, 30–33
  uncovering hidden patterns of behavior of, 101–104
Customer-centric paradigm, creating a, 17–18
Customer journey mapping, 104–111
  elements of, 104–105
  and predicting intent, 106–108
  and predicting returns, 108–111

Index

D

DALL•E2 tool, 87
Dashboards, 8
Data analytics, 18
Data Center for Excellence, 6
Data-centric AI approach, 58
Data-driven decision-making, 6–8
Data operations (DataOps), 157
Datasets, choosing the right, 7–9
Da Vinci, Leonardo, 155
Decision-making:
   in behavioral economics, 28–30
   data-driven, 6–8
   dual-process theories of, 33–35
   effect of "noise" on, 39–40
   heuristics and biases in, 22, 35–39
   human, 3
   and nudging, 42–46
   online, 26
   in prospect theory, 40–42
   psychology of, 21–22, 33–42
   "rational," 33–34
   with rule-based engines vs. ML algorithms, 10–11
Decision trees, 81
Decoy pricing, 44–45
Deep Blue, 77
Deep learning, 82–86
Demand forecasting models, 63, 127–131
Demand-sensing models, 62–63
Descriptive statistics, 16
Digital entertainment services, 15
Digital identity (DI), 69–71
Digital platforms, rise of, 26
Discounts, 12–13
Doudna, Jennifer, 175
Dual-process theories of decision-making, 33–35
Dual process theory, 40
Dumb Money (film), 98–99
Dunning-Kruger effect, 38
Dynamic surge pricing, 49

E

Eating habits, encouraging better, 24
E-commerce data sources, 106
Economics, 5
Economic theory, 3
"Edmond de Belamy" (AI portrait), 88
Efficient market hypothesis, 25
Einstein, Albert, xi, 137, 175
Emotions, 12, 20
Endowment effect, 38, 41–42
Equity premium puzzle, 25
Existence bias, 143
Expected utility theory (EUT), 40–42
Experimentation (behavioral economics), 46–51
   as game changer, 97
   goal of, 47
   process of, 48–49
   and scaling, 169–172
   by Uber, 49–51
Expert systems, 77

F

Fashion industry, 88
Feature engineering, 79
Feedback, 17
   instant, 49
   in reinforcement learning, 80
Feynman, Richard P., 1, 46, 175
Field experiments, 48
Field-programmable gate arrays (FPGAs), 56
Filmmaking industry, 87
Filtering, data, 113
Finance industry, 122
First African School of Physics, xiii
First impressions, 34
Fisher, Irving, 46
Forecasting, 127–135
   and dynamics of demand, 127–131

Forecasting (*continued*)
   with ensembling models, 133
   and inventory
      optimization, 133–135
   with transparent mixed marketing
      models, 132–133
Forecasting, accuracy of, 9
Fortune 500 companies, xviii, 3–4
401(k)s, 31
Framing effects, 22, 23, 38, 42
Fraud detection, 86
Functional magnetic resonance
      imaging (fMRI), 29

G
GameStop, 98–99
Game theory, xv–xvi, 5
Gaming industry, 87–88, 92
Gartner, 74, 89, 90, 157
The Geek Way (McAfee), 175
General Data Protection
      Regulation, 89
*The General Theory of Employment,
      Interest, and Money*
      (Keynes), 21
Generative adversarial networks
      (GANs), 85–88
Generative AI, 86–90
Gen Z, 31, 67
Go (game), 92
Goal setting, 49
Google Assistant, 84, 157
Gradient boosting machines
      (GBMs), 82
Grocery shopping, online, 16, 107
Grocery stores, 45–46, 116–117
Group theory, xiii
Gut instincts, relying on, 8

H
Halo effect, 38
Harvard Business Review, 50
Health care, 27, 64–65, 77, 84, 85

Hedge funds, 99
Heuristics, 22, 33, 35–39
Higgs boson, xviii, 53
Hindsight bias, 38
Hollywood, 87
Homo Economicus, 20, 24
Huang, Jensen, 95
Human decision-making,
      3, 12
"Humans," 20, 24
Hypotheses, 48, 56

I
IBM, 77, 93
Impact bug, xiii, xiv
Indigo bookstores, 13
Inferential statistics, 16
Influencers, 13
The Innovator's Dilemma
      (Christensen), 175
Instagram, 10
Instant feedback, 49
In-store shopping, recommendation
      models for enhancement
      of, 116–118
Insurance, 31, 41
Intellectual diversity, 159–164
Intent, predicting customer,
      106–108
Internet of Things, 93
Interpretation layers, 166–169
Intuit, 51
Intuition, developing, 165
Inventory:
   optimization of, 133–135
   reduced excess, 9
IRAs, 31

J
Jackson, Matthew, xv
Jimmy Choo shoes, 126
Journal of Behavioral Economics,
      23–24

K

Kahneman, Daniel, 19, 22–23, 35, 39–40, 42
Kala, Devkant, 28
Kasparov, Garry, 77
Katona, George, 21–22
Keynes, John Maynard, 21
Key performance indicators (KPIs), 110
K-nearest neighbors (KNN), 81

L

Lab experiments, 48
Life insurance, 31
Linear regression, 81
LinkedIn, 19, 30, 74
Loan applications, 79
Local purchasing, 15–16
Lockdowns, 15
Loewenstein, George, 144
Longevity bias, 143
Long short-term memory (LSTM) networks, 84
Loonshots (Bahcall), xiv
Loss aversion, 13, 14, 22, 23, 25, 37, 43, 49, 103, 140–145
Lottery tickets, 41

M

Machine learning (ML), 1–3, 9
  about, 78–82
  algorithms, 10–12
  biologically inspired models for, 90, 92–94
  deep learning as subset of, 82–86
  iterative nature of, 80
  and minimum viable data, 58
  overview of, 75*f*
  prediction using, 16–18
  and survey data, 67–71
Mapping customer journeys, *see* Customer journey mapping

Marketing, 50, 60, 67, 81–82, 88–89, 105, 119, 121
Mathematics, xii–xiii, 5, 34
Medical applications, *see* Health care
Medication adherence, 27–28
Memphis, Tenn., 60
Mental accounting, 25
Microsoft, 157
Millennials, 67
Minimal viable product (MVP), 58–59
Minimum viable data (MVD), xviii, 7, 53–71
  and asking the right questions, 59–63
  and concept of minimal viable product, 58–59
  with large amounts of data, 55–58
  for retailers, 102
  and survey data, 67–71
  and synthetic data, 63–67
Mixed marketing models (MMMs), 131–133
Mock data, 65–66
Model operations (ModelOps), 157
Monte Carlo data, 66–67
Monte Carlo simulations, 66–67
Morningstar, 114
Mortgages, 31, 120–121
Multidisciplinarity, xvii, xviii, 159–160, 162
Music production, 85
MYCIN, 77

N

National Retail Federation, 108
Natural language processing, 17, 84
Natural selection, 90
Neoclassical economics, 20
Netflix, 16–17
Neural networks (NNs), 90, 106–108, 113, 116, 134
Neuroeconomics, 29

Neuromorphic computing, 93
New York City, 114
Ng, Andrew, 58
Noise, 39–40
Noise: A Flaw in Human Judgment (Sunstein and Sibony), 39
Not sufficient funds (NSF), 122–123
Nudge (Thaler and Sunstein), 24, 127
Nudging, 24–26, 42–46

O
Obstacle avoidance, 93
Online behavior, 26, 106
On time and in full (OTIF) orders/deliveries, 139, 140
OpenAI, 87
Oppenheimer, J. Robert, 175
Opt-in systems, opt-out vs., 43
Organ donation, 43
Overconfidence bias, 38

P
Pareto, Vilfredo, 46
Partially synthetic data, 64–65
Particle physics, xiii–xvi, 53–54
Past experiences, 12
Path planning, 93
Patterns, uncovering hidden, 10–11, 101–104
Paying, pain of, 29
Pension plans, 24
Perfect competition, 46
Personalization, 18
Personalized pricing, 121–127
Peterson, Richard L., 29
Point of sale (POS) data sources, 106, 116, 128, 129
Positive price elasticity, 7–8
Predictably Irrational (Ariely), 26–27
Preferences, customer, 15
Price elasticity, 6–9
Pricing:
  decoy, 44–45
  dynamic surge, 49
  personalized, 121–127
Procrastination, 25
Prolog, 77–78
Propensity models, 118–121
Prospect theory, 23, 40–42
Prospect Theory (Kahneman and Tversky), 22
Psychology, 3, 21–22, 26. *See also* Bias(es)
Public health, 27
Public policy, 24

Q
Questions, asking the right, 4, 5, 57, 59–63
Questionnaires, 68

R
Ralph Lauren, 126
Random assignment, 48
Randomized control trials (RCTs), 170
Recommendation models, 16–17, 111–118
  effectiveness of, 112
  enhancement of in-store experience with, 116–118
  goal of, 112
  and nuanced customer preferences, 114–116
Recurrent neural networks (RNNs), 84–85, 115
Reddit, 98–99
Reinforcement learning (RL), 80, 92
Representativeness heuristic, 22, 36
Retirement savings, 24, 25, 31, 43
Returns, product, 103, 108–111
Return on investment (ROI), 61, 62, 74, 131–133, 171–172
Reviews, 17

Index

Risk aversion, 41, 103, 139, 141, 151
Robinson, Alan, 77
Robotics, 80, 92, 93
Rule-based engines, 10–12

S

"Save More Tomorrow" (SMarT) program, 25, 43
Savings accounts (savings rates), 11–12, 25
Scaling of AI projects, 155–174
   and building trust, 164–166
   and effective communication, 159–161
   and experimentation, 169–172
   and intellectual diversity, 159–164
   and interpretation layers, 166–169
   success factors in, 156–159
Season patterns, 15
Segmentation, 97–104
Self-selection bias, 32
Sherlock Holmes, 53
Shopping experiences, safer, 15
Sibony, Olivier, 39
Silicon Valley, xvii
Simon, Herbert, 21
Siri, 82
SKU codes, 9, 113, 134
"Smart sizing," 58
Smith, Vernon, 46–47
Social and Economic Networks (Jackson), xv
Social media, 13, 98–99
Social proof, 13
Social status, 21
Spam emails, detection of, 81
Speech recognition systems, 84
S-shaped diffusion curve, 70
Stanford University, xvii
StarCraft II, 92
Statistics, 16
Statistical analysis, 67

Status quo bias, 38, 43, 140–145, 150
Stock market prices, 25
Stockouts, 9
Strategic placement, 45
String theory, xii–xiii
Student loans, 31
Substitute products, recommended, 105
Success, enablers of, 156–159
Sunk cost fallacy, 38, 148–149
Sunstein, Cass R., 24, 39
Supply chains, 15, 63, 93, 101–102, 127, 137, 142
Support vector machines (SVMs), 81
Surge pricing, dynamic, 49
Surveys (survey data), 32, 67–71
Survival rates, 122, 125–126
Swarm intelligence, 90, 93
Symbolic AI, 76
Synthetic data, 63–67
System 1 responses, System 2 vs., 34–35

T

Tabriz, Iran, xii
Temporal dynamics, 15
Tesla, 84–85
Testimonials, 13
Text message reminders, 27
Thaler, Richard, 20, 24, 25, 42–43, 99, 127
The Theory of the Leisure Class (Veblen), 21
Theory+Practice, xvii–xviii, 60
Transformers, 85–86
Transparency, 11, 13, 132–133
Trend cycles, 15
TrueNorth chip, 93
Trust, 13, 89, 164–166
Tversky, Amos, 22, 23, 35, 40, 42

U
Uber, 49–51
United Kingdom, 24–25, 116–117

V
Vancouver, B.C., xii, xvii
Variational autoencoders (VAEs), 86
Veblen, Thorstein, 21
Virtual assistants, 82, 84, 85

W
Waymo, 84–85
Weber, Martin, 144
Work-from-home, 15

X
XGBoost, 9, 106, 107, 118–119, 123, 129
X-rays, 84